Yajnaseni Dash

Neural Networks Boost Software Maintainability Analysis

AF180323

Yajnaseni Dash

Neural Networks Boost Software Maintainability Analysis

Leveraging AI to Assess and Enhance Object-Oriented System Maintenance

LAP LAMBERT Academic Publishing

Imprint
Any brand names and product names mentioned in this book are subject to trademark, brand or patent protection and are trademarks or registered trademarks of their respective holders. The use of brand names, product names, common names, trade names, product descriptions etc. even without a particular marking in this work is in no way to be construed to mean that such names may be regarded as unrestricted in respect of trademark and brand protection legislation and could thus be used by anyone.

Cover image: www.ingimage.com

Publisher:
LAP LAMBERT Academic Publishing
is a trademark of
Dodo Books Indian Ocean Ltd. and OmniScriptum S.R.L publishing group

120 High Road, East Finchley, London, N2 9ED, United Kingdom
Str. Armeneasca 28/1, office 1, Chisinau MD-2012, Republic of Moldova, Europe
Managing Directors: Ieva Konstantinova, Victoria Ursu
info@omniscriptum.com

Printed at: see last page
ISBN: 978-620-8-44338-2

Copyright © Yajnaseni Dash
Copyright © 2025 Dodo Books Indian Ocean Ltd. and OmniScriptum S.R.L publishing group

ABSTRACT

Software quality is an integral aspect of system development that determines the degree to which the software is in use will meet the requirements of the customer. Maintainability assessment is an essential aspect of software development. However, it is a cumbersome process. Many methodologies are proposed so far to estimate the maintainability of software. The relationship between OO metrics and software maintenance effort is complex and nonlinear. Since the OO paradigm is different from the procedural paradigm, different software design metrics have to be defined and used. In this study, surveys have been conducted on the different works regarding software maintainability on the object-oriented paradigm. Artificial neural network is one of the sophisticated techniques that have a great predictive capability. This paper investigated the application of neural networks to predict maintainability of the software using object-oriented metrics. In this study maintenance effort was chosen as the dependent variable and principal components of object-oriented metrics as the independent variables. Maintainability is predicted using Multi-Layer Perceptron (MLP) neural network. The results of the study are also compared with other models, and it is revealed that the model presented is more useful than the previous one.

CONTENTS

LIST OF FIGURES

LIST OF TABLES

LIST OF ABBREVIATIONS

ANN- Artificial Neural Network

C&K- Chidamber & Kemerer

DAC- Data Abstraction Coupling

DIT - Depth of Inheritance Tree

GRNN- General Regression Neural Network

LADY- Language for Distributed Systems to Study Maintainability

LCOM - Lack of Cohesion in Methods

MAE- Mean Absolute Error

MLP – Multi Layer Perceptron

MPC- Message Passing Coupling

MTTR - Mean Time To Repair

NOC - Number of Children

NOM- Number of Local Methods

OO- Object Oriented

PCA- Principal Component Analysis

QUES-Quality Evaluation system

RFC -Response For a Class

RMSE -Root Mean Square Error

SPSS- Statistical Package for the Social Sciences

UIMS -User Interface System

WMC- Weighted Methods per Class

Chapter 1

INTRODUCTION

Software quality is an integral aspect of the development scheme that determines the degree to which the software in use will meet the expectations of the customer. Currently software quality is a major factor of concern. The growing research activity in software quality leads to innovation of novel techniques, to predict its attributes. Maintainability is an important quality attribute and a difficult concept as it involves many measurements. Object-oriented (OO) metrics are used in quality estimation. However, quality estimation means estimating maintainability or reliability of software. Software reliability is a valuable ingredient to make the system work properly without a fail [1]. As the OO system uses a huge number of small methods, it is time consuming, error prone and has a distinctive maintenance problem [2]. Artificial Neural Network (ANN) is one such technique which involves a series of steps for the computation of maintainability effort.

As traditional computers are not excellent for interacting because of noisy data, immense parallelism, fault tolerance, and failure to adapt to certain circumstances, so ANN provides a better option for handling software quality. The neural network system also helps us when we are unable to devise an algorithmic elucidation. The application of ANN for software maintainability prediction using object-oriented metrics is focused in this thesis.

1.1 OBJECT-ORIENTED SOFTWARE SYSTEMS

Some of the fundamental terms and concepts that are associated with object-oriented software systems are presented in this section.

According to Cox, the object-oriented pattern is “a system building tool which puts the reusability of software at the centre of the software development process” [3]. Wegner also explored that object-oriented system consists of objects, classes, inheritance [4].

A class refers the pattern of objects and is the crucial system constituent in the object-oriented paradigm. Objects that share the similar definitions are assembled into classes. An instance of a class is called as an object. Inheritance and polymorphism are the two unique features to object-oriented systems. Inheritance is a reuse method which lets programmers for defining the classes incrementally. This can be achieved by reusing formerly defined classes as the base for latest classes. Polymorphism is the ability of classes to possess several forms. This mechanism can perform replacement of an object with its sub objects. The principal objective of using these concepts is to reduce the OO system complexity by improving software quality.

Generally, it is very difficult to accomplish a good software design because of its error prone nature. The presence of defects in software compromises its quality. According to McCall's [5] there are 11 quality attributes and among which maintainability has its own significance. However, maintainability is the costliest activity in the whole lifecycle of software development and much effort is needed to complete this phase. Design errors during development have a negative impact on maintainability. Hence identification of these blemishes and solving these issues is an essential concern for enhancing the quality of the software.

Object-oriented (OO) systems are also associated with bugs. So production of good software by a large-scale legacy system written in an OO language is a cumbersome process. Although the syntaxes and the concepts of this language are known, these legacy systems are monolithic, not flexible, and difficult to maintain. Either a perfect OO system follows the rules and design heuristics or requires proper quantification for controlling the quality. De Marco [6] justified that good design rules cannot be expressed in a quantifiable manner. Assessing maintainability accurately not only helps the developer to improve the design and coding but also increases the performance of the system by reducing the complexity. Hence, estimating maintainability with novel techniques leads to enhancement of quality of the software system.

1.2 SOFTWARE MAINTENANCE

Maintenance is the act of keeping an entity in an accessible state of repair or validating it to protect from failures. Maintenance process is needed to modify the software product after the delivery procedure for rectification of the existing faults, enhancement of the system performance, and adaptation of the software to the newer environment. Boehm [7] suggested that 3 steps are carried out during the process of maintenance. (Figure 1.2).

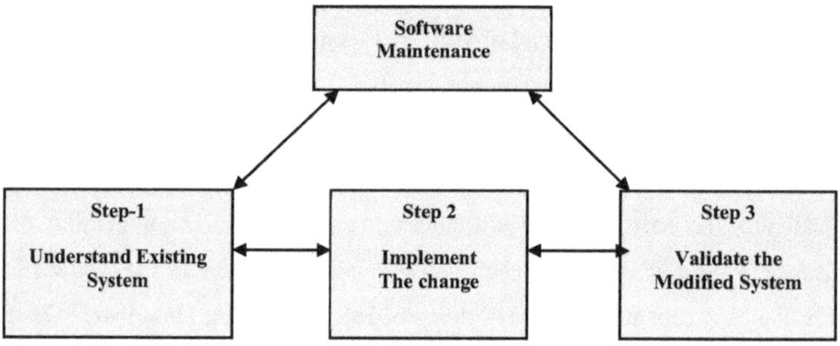

Figure 1.2: Steps of Maintenance

1.1.1 Types of Maintenance

Many researchers classified software maintenance in terms of adaptive, corrective, perfective [8] and preventive [9].

a. Adaptive Maintenance

The necessity of using this environment-driven maintenance occurs if there are alterations in hardware, operating systems, files, or compilers, which have an impact on the system.

b. Corrective Maintenance

This error-driven activity is like the debugging process and takes place after the system is placed in operation. Corrective maintenance is a requisite during the complete life cycle of a system as software programs are error prone. The bulk of maintenance behaviors are of this kind.

c. Preventative Maintenance

It is used to make the software more maintainable by updating the documents. The changes are done for detection and correction of latent faults in the software product.

According to recent studies, 90% of the maintenance is carried out in either corrective or perfective ways whereas corrective alone corresponds to 70% of all modifications [10]. Thus, corrective maintenance is known as 'traditional maintenance' and the other forms are referred to as 'software evolution'.

1.3 MAINTAINABILITY

Maintainability is the ease with which the process of maintenance is carried out. Earlier, it can be defined in two ways, either informally or as a function of directly measurable attributes. According to the IEEE standard glossary of software engineering terminology, maintainability is defined as "the ease with which a software system or component can be modified to correct faults, augment performance and adapt to a changed environment" [11].

The other way of defining maintainability is "the capability of the software product to be modified". The modification is in terms of corrections and improvements of the software to changes in environment, and in requirements and functional specifications [12].

1.3.1 Measuring of Maintainability

Regardless of the immanency of any exertion for measuring maintainability, immense effort has been made to construct formulas for depicting maintainability. Maintainability can be described as a function of instantly computable attributes from A1 to An, such as, $M = f(A1, A2, \ldots, An)$. Informally, this technique is somewhat interesting, as it is instinctive that a maintainable system must be simple and reliable. Conversely, there may be complications to measure the attributes, to weigh them against each other and to merge them in a function f. However, such an aspiration is relatively inadequate to certain contexts, namely a kind of system, a category of project, a programming language, the skill and knowledge of people concerned for concluding. According to IEEE-1219 [13] explicit measures of software maintenance are analyzability, changeability, stability, and testability.

1.3.1.1 Software Metrics

The need for measurement arises not only to reduce costs, effort, and schedule but also to amplify the system performance. Certain measures of the maintainability of software can be obtained using available commercial tools. However, maintainability can be estimated by applying software metrics. Software metrics is defined as, "The continuous application of measurement-based techniques to the software development process and its products to supply meaningful and timely management information, together with the use of those techniques to improve that process and its product" [14].

1.3.1.2 Usage of Software Metrics in Measuring Maintainability

Various studies have been carried out to enumerate maintainability by using many kinds of measures. Software maturity index is based on measuring the stability of the product. Effort metrics are commonly used to estimate the maintenance effort by calculating the Mean Time To Change (MTTC) [15]. Syntactic complexity family of metrics, statically analyzing source code to measure maintenance effort, and constitutes Halstead Volume (HV) metric [16] and McCabe's cyclomatic complexity (CC) [17]. HV, in turn, is a combined metric based upon a distinct number of operators and operands present in source code. Among all probably the most appreciable is the Maintainability Index, MI [18-19]. They have mainly determined the maintainability of the software which is based upon the status of the subsequent source code. This index is a complex number that depends on different metrics of a software system. MI is based upon the HV metric [16], the CC metric [17], lines of code metric (LOC), and possibly the comment lines per module's percentage (COM). Software maturity index is based on measuring the stability of the product [20].

The metrics used to measure maintainability, and their drawbacks are described in Table 1.3.1.2.

Table 1.3.1.2: Measuring metric

Measuring Metrics	*Author*	*Year*	*Reference*	*Description*	*Drawbacks*
McCabe's cyclomatic complexity number	Mccabe	1976	[17]	Graphically measure the number of independent	It is impossible to specify the total paths.

				paths in a program. V(G) = E – N + 2P	Higher cyclomatic complexity V(G) has a negative impact on changing the system.
Halsted volume	M.H. Halsted	1977	[16]	Used to predict maintenance effort and mean number of faults in programs.	Between the lexical complexities of code, there exist weak logical associations.
Effort metrics	Pressman	1982	[15]	The lower value of Mean Time To Change (MTTC) leads to more maintainable product.	Lack of prediction capability and dependency on skill of maintainer.
Maintainability Index (MI)	Oman et al, and Coleman et al	1992, 1994	[18-19]	Derivation is from source code. Effectively analyzes many systems by making MI comparisons.	Identification of high-risk modules provides outstanding insight into the source code. Hence there is no major drawback.
Software Maturity Index (SMI)	IEEE 982.1	1998, 2005	[20]	If SMI becomes 1.0 then the product does not require modifications and stabilized. $SMI = (M_T - (F_a + F_c + F_d))/M_T$	Module measurement is not associated with product stability and difficult interpretation of negative values.

1.4 ARTIFICIAL NEURAL NETWORK

The Neural network is a network of interconnected neurons where information propagation occurs by firing electrical pulses via its connections. During the lifetime of a neuron, the connections or weights need to be adjusted. Similarly, the quantity of incoming pulses which are required in activating a neuron is changed. These are the behavioral attributes that let the NN learn. Generally, there are 3 basic learning algorithms in neural networks, such as supervised learning, unsupervised learning, and reinforcement learning. Supervised learning is most used and also called learning with a teacher. It is applied when the target value is known and associated with each input in the training set.

1.4.1 History of Neural Network

An ANN is an information processing paradigm that is an emulation of a biological neural system. A trained neural network can be thought of as an expert in the category of information it has been given to analyze.

The first essential model of neural network was proposed by McCulloch and Pitts [21]. It is a computing model of the activities of the nervous system and acts as a binary device where every neuron has predetermined threshold logic. Various researchers (e.g. John von Neumann, Marvin Minsky, Frank Rosenblatt) worked on this model.

Hebb [22] described in his standard book "The Organization of Behaviour", the proper interconnection and self-organization of neurons. The existing path, which lies between the neurons, strengthens the connections. As an enormous number of simple neurons embedded in an interactive nervous system, it is promising to provide computational power for extremely complicated information processing. The basic learning rule in neural network literature is Hebb's law. The perceptron model was developed by Rosenblatt [23], which follows the perceptron learning law. ADALINE (Adaptive Linear Element) for computing element and LMS (Least mean square) learning algorithms explained by Windrows and Hoff [24] for adjusting the weights of an ADALINE model. Rumelhart and McClelland [25] illustrated the weight of multilayer feed forward neurons can be adjusted in a logical approach by mapping them implicitly in a set of input and output pattern pairs for learning purposes. This act of learning is named as the generalized delta rule or error back propagation rule. A marvelous outline of ANN was suggested by Cheng and Titterington [26] and Warner and Mishra [27]. Zang et al. [28] provided the broad-spectrum procedures for network modeling for the use of forecasting purposes by using statistical techniques. The performance of multilayer perception and linear regression models for prediction of quality was also compared. There are various commercial and free software available that have been made for the prediction of faults in the neural networks.

1.4.2 Neural Network Architecture

Neural network architecture refers to the types of interconnections between neurons. A network is said to be fully connected if the output from a neuron is connected to every other neuron in the next layer [29].

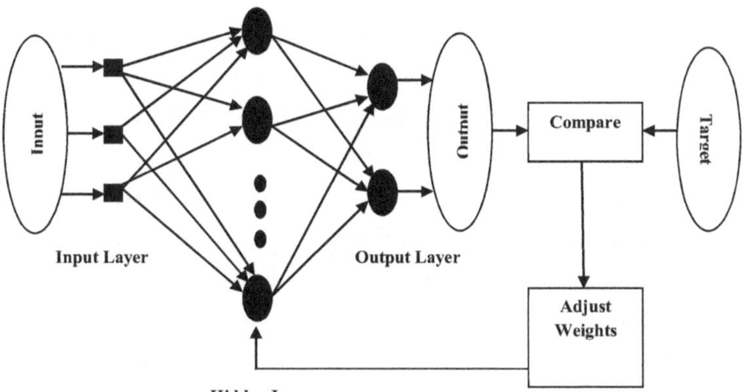

Figure 1.4.2: Sample architecture of ANN

There are three layers in an ANN, such as the input layer, the hidden layer, and the output layer as shown in Figure 1.4.2.

1) Input Layer: This layer comprises the input units, which symbolize the unrefined information provided for the network.

2) Hidden Layer: This layer is represented by hidden units which are influenced by the behavior of the input units and the weights that connect this input and the hidden units.

3) Output Layer: The output unit's behavior is dependent on the specificity of the hidden units and the weights connecting the hidden and output units.

Generally, the ANN adjusts the values of the weights to produce a specific target output from a particular input. Then it compares the output with the target till a match between the target and the output is found. A neural network can be trained to do a specific function only by considering several such input or target pairs.

1.4.3 Connection Structures of Neural Network

A neural network comprises neurons and weight building blocks. The network activities mainly depend upon the communications between these building blocks. There are 4 kinds of connections namely feed forward, feedback, lateral, and time-delayed connections [30-31].

1) Feed forward connections: Data is transmitted from lower layer to upper layer by feed forward neural networks. It is a one-way communication process where the signal flows from input to output. Multilayer feed forward network is also known as multilayer perception.

2) Feedback Connections: Data is brought from the higher layer to the lower layer by feedback networks. It is a two-way communication process where signal flows in a bidirectional manner. Symmetric recurrent neural networks are also known as attractors or Hopfield neural networks.

3) Lateral Connections: One typical instance of the lateral connection is the winners-take-all circuit. It is used to confer an essential role to select the winner. In the map example, by allowing neurons to interact via the lateral network, a definite network topology ordering relationship can be preserved.

4) Time-delayed Connections: Temporal dynamics models can be produced by incorporating delay elements. Mostly these connections are appropriate for temporal pattern recognition.

1.4.4 ANN based Measurement of Maintainability Effort

Computation of software effort is a challenging task as the raw data available during initial parts of the project development are inconsistent, haphazard and incomplete. All the estimation models are greatly used to predict the software effort based on the mathematical formula of effort such as effort = $\alpha \times$ size β. Besides this, various techniques like ANN, analogy-based reasoning, regression trees, rule-based induction models are also considered for the estimation of software effort. Idri et al. [32] explored cost estimation models based on ANN. They had used back-propagation as an algorithm and feed forward multilayer perceptron as architecture. The sigmoid function was selected by them as the activation function of ANN. Various studies have been performed to estimate software effort accurately by applying NN approach. Prasad et al. [33] compared software effort estimation by using radial basis and generalized regression neural networks and found that radial basis neural network is more suitable.

1.4.4.1 Advantages and Disadvantages

Use of ANN for prediction of software effort possesses several advantages; still, there are causes of not preferring the ANN technique for software cost estimation. Some of these are presented above in Table 1.4.4.1.

Table 1.4.4.1: Pros and Cons of ANN in Effort Prediction

S.No.	Advantages	Disadvantages
1	Software development is an incessantly emerging process. 'Learning from the past situations' is the most important criterion of ANN, which acts like a path finder for future development of software.	NN approach is just like the black box technique where we are unable to understand the internal structure of the network.
2	It is applicable for developing efficient algorithms in the fields where little knowledge is available.	It can solve problems comprising high complexity, such as classification and categorization type of problems. However, in case of cost estimation, generalization is taken into consideration rather than classification.
3	It can derive precious information and regularity from large databases inherently.	No guidelines are available to construct the NN topologies.
4	It can work efficiently where large numbers of variables are present and the conditions are changing frequently.	Decision making in ANN is a difficult task, still it is used by researchers for its learning mechanism.
5	ANN can correlate composite relationship between dependent variable (e.g. maintenance effort) and independent variable (e.g. software metrics).	The advantages of ANN suppress its disadvantages.

1.5 SUMMARY

Several measures have been proposed by researchers to capture the quality of OO code and design and used for identifying fault-proneness of classes. Various investigations have also been made to predict software quality by using statistical methods. ANN has seen an explosion of interest over the years and is being successfully applied across a range of problem domains such as finance, medicine, engineering, geology and physics. Indeed, if there are problems of prediction, classification or control, neural networks will be an efficient asset to solve them.

Chapter 2

LITERATURE SURVEY

2.1 INTRODUCTION

Various researchers have proposed models of software quality attributes that can be useful for discussing, planning, and rating the quality of software products. The models often include metrics to measure the degree to which the software attains each quality characteristic. The various attributes of quality are namely maintainability, usability, efficiency, reliability, probability, and functionality. Among these attributes, maintainability has its importance due to its consequences. This chapter surveys the different studies regarding software maintainability on the object-oriented paradigm which provide further assistance in succeeding research. It also highlights various research methodologies proposed to predict the quality of OO metrics by using neural network approach.

2.2 LITERATURE SURVEY OF MAINTAINABILITY AND OBJECT-ORIENTED SYSTEM

Rombach [35] reported the outcome of a controlled experiment for studying the distributed software's maintainability. They have experimented with a partial object-oriented language LADY, a Language for Distributed systems. He summarized the findings by concluding that software that was written in an OO language is much more maintainable than software written in a conventional language. The instigator also explored that complexity metrics might estimate maintenance and understand the consequent code measures. Johnson and Foote found the use of numerous small methods, which influence excellent programming style of object-oriented system [2]. According to Moreau [36-38] conventional metrics are not suitable for OO systems for many reasons. Firstly, the hypothesis related to program size and programmer productivity in structured systems need not be applied directly to OO systems. Secondly, the conventional metrics do not accomplish the structural features of OO systems. Thirdly, the calculation of complexity of the total system, which is the same as the summation of the complexity of its components, is not proper for OO systems. Moreau [37] also stated that the traditional software metrics, which are existing, might be efficient in a specific method within an object. However, there is no empirical evidence present to support these statements.

Moreau [38] again employed traditional metrics to compare two implementations of a graphics editor. One of them is a traditional implementation in C, and the other one is the object-oriented implementation in C++. Mancl and Havanas [39] presented a case study of the impact of the C++ programming language and object-oriented design on the maintenance state of a software development project. The result had shown that while using the object-oriented design, there is variation in terms of improvement in reusability and decrease in complexity of software. This study attempted to assess the differences between the two types of programming, i.e., object-oriented ones and conventional structured one. The measurements identified some of the key points where object-oriented programming had a significant effect on augmenting productivity. They have applied "the number of lines of code modified per maintenance task" as a measure of the maintenance effort. Inheritance and polymorphism methods not only strengthen object-oriented systems but also make the process of program understanding and analysis difficult. Wilde and Huitt [40] described several maintenance issues which are exclusive to object-oriented systems. They have taken instances from a PC smalltalk environment and two real-world systems. They also have analyzed difficulties associated with object dependencies, dynamic binding, control of polymorphism, dispersed program structure, high-level understanding and detailed code understanding. Chidamber & Kemerer (C&K) also explored some of the object-oriented design metrics for computing maintainability effort. These metrics include WMC, DIT, NOC, CBO, RFC and LCOM [41-43]. Li and Henry [44] have analyzed the applicability of object-oriented complexity metrics proposed by C&K in the field of maintenance in two commercial systems. According to their conclusion, these metrics in general can be utilized as an interpreter of maintenance effort; however, two of these metrics have not shown the expected result.

C&K [41-43] experimentally analyzed the OO design metrics to evaluate the utility of these metrics for assessing the probability of fault detection among classes. It is obvious from the description that the definitions of these metrics are somewhat dependent upon language. Basili et al. [45] as a consequence, had slightly adjusted some of C&K's metrics to reflect the specificities of the C++ language. Although, the weakness related to this study can be the use of 'the number of lines changed per class as a measure of the maintenance effort. Li and Henry [44] also defined some additional metrics. Abreu and Melo [46] also experimentally studied that they have obtained their MOOD metrics for correlating with the system maintenance and reliability. Harrison et al. [47] defined the usefulness of a design metric (the number of test cases) to predict the testing time. Binkley and Schach [48] applied class coupling methods to validate the coupling dependency metrics as a predictor of run time failures and maintainability measures by the usage of the C++ system (patient core management), 113 classes, 82KLOC, file transfer facility, 29 java classes and 6 KLOC measures. Tang et al. [49] empirically studied object-oriented metrics by using C & K metrics, but they have not considered the LCOM and LOC methods as variables.

Muthanna et al. [50] investigated the use of software design metrics to statistically estimate the maintainability of large software systems, and to identify error prone modules. It is a reality that the application of metrics in the analysis and design of object-oriented software systems can assist designers in making better decisions in gaining relevance in the software measurement arena. Polo et al. [51] reported code metrics for prediction of maintenance of legacy programs as a case study by applying logistic regression, MANTEMA, a methodology for maintenance. Moreover, the requirement of having early predictors of external quality characteristics depends on maintainability. In addition to this, the aim is to explore the utility of early metrics as early class diagram maintainability indicators. It is to be noted that early metrics refer to those which estimate internal characters, such as structural complexity and size of UML class diagrams. For this purpose, Genero et al. [52] presented a controlled experiment and its replication, which was carried out to gather the empirical data. Subramanyam and Krishnan [53] conducted an experimental analysis on a subset of C & K metrics suite in determining software defects. Hayes et al. [54] proposed a metrics-based software maintenance effort model by the use of validated datasets. They applied COCOMO (Constructive Cost Estimation Model), SLIM, AMEffMo(Adaptive Maintenance Effort Model) and Regression Analysis methods for computing the effort. Kiewkanya et al. [55] constructed a maintainability model of object-oriented design by applying methods like association, aggregation, generalization and classification.

Alain et al. [56] proposed prediction of maintainability by the use of regression analysis and DC ratio methods. Bocco et al. [57] assessed the capacity of internal metrics as early predictors of maintenance effort through different experiments. The utility of measures for the analysis and design of object-oriented systems is widely being identified in the field of software engineering research. Recognition of the need for early indicators of external quality attributes is increasing. Genero et al. [58] investigated through vivid research whether a group of UML class diagram measures could be excellent estimators of two main sub-distinctiveness of the maintainability of class diagrams, namely modifiability and understandability. Results gathered from a controlled experiment support valuable estimation models for building these features as an early measure. Particularly, the measures were capturing structural convolution through generalizations and associations. Likewise, these measures have been associated with the subjective perception of the subjects about the convolution of the diagrams. This fact showed that to some extent, objective estimation captures similar features as the subjective ones. Breesam [59] empirically formalized a set of metrics to use them for the quality measurement of an OO design with consideration of class inheritance. Neelamegan and Punithavalli [60] surveyed four OO quality metrics, and they found that these metrics focus on measurements when applied to the class and design characteristics.

Sastry and Saradhi [61] attempted to implement software metrics with assistance of GUI and also examined relationships of metrics for determination of quality. The quantity of software attributes estimated by observing the life cycle of object-oriented software development. Amjan Shaik et al. [62] have statistically analyzed the OO software metrics on C&K metric collections by verifying the data put together from the projects of some students. Fast feedback for software designers and managers was provided by applying metrics data. They examined minutely that the proper use of these metrics directly reduces the cost of the overall execution and enhances the quality of the ultimate product. Rizvi and Khan [63] proposed MEMOOD model that caused an improvement of the maintainability of class diagram and also enhanced the maintainability in the final software. Thapaliyal and Garima [64] have empirically analyzed the software defects and object-oriented metrics. They evaluated two metrics, the weighted method per class (WMC) and the coupling between object classes (CBO) of the C&K metrics suite. They investigated whether the metrics taken are linked to defects or not by the usage of 50 samples of Java classes of different projects. Dubey and Rana [65] used a metrics approach to assess the utility of object-oriented software to develop successful software applications. Gautam and Kang [66] described that the compound MEMOOD model is better to determine the maintainability of class diagrams in terms of their scalability, understandability, modifiability and level of complexity. Malhotra and Jain [67] reviewed software fault prediction for OO systems and used logistic regression method. Shaik et al. [68] suggested a metric approach for evaluating the system test cases in an OO system. They have used the regression CR tool architecture method for the estimation process. The detailed survey is summarized in Table 2.2.

Table 2.2: Studies on Maintainability and Object-Oriented Systems

Sr. No.	*Author*	*Year*	*Description*
1	Rombach	1987	Controlled experiment on LADY, a Language for Distributed systems to study maintainability. Summarized that the software written in object-oriented language is more maintainable than software written in a conventional language.
2	Johnson and Foote	1988	It is found that the use of numerous small methods seems to influence an excellent programming style of an object-oriented system.
3	Moreau	1987, 1989, 1990	The unsuitability of traditional metrics for OO systems is due to: a) Program size and programmer productivity in structured systems need not be applied directly to OO systems. b) Conventional metrics do not accomplish the structural features of OO systems. c) Calculation of complexity of the system, which is the same as the summation of the complexity of the components, is not proper for OO systems.
4	Mancl and Havanas	1990	A case study of the impact of maintainability on OO design. Applied 'the number of lines of code modified per maintenance task' as a measure of the maintenance effort. The results showed improved software reuse and decreased complexity of software in the parts of the project to use the object-oriented design.
5	Wilde and Huitt	1992	Described several maintenance issues which are unique to OO systems. Analyzed difficulties associated with dynamic binding, object dependencies, dispersed program structure, control of polymorphism, high-level understanding, and detailed code understanding.
6	Li and Henry	1993	Authors studied the applicability of OO complexity metrics with reference to the maintenance in two commercial systems and conclude that these metrics in general can be used as predictors of maintenance effort,

			though two of the metrics were not as good as expected.
7	Chidamber and Kemerer	1994	Explored some of the OO design metrics for computing maintainability effort. These metrics include WMC, DIT, NOC, CBO, RFC and LCOM
8	Chidamber and Kemerer	1995	Authors evaluated whether or not the above metrics are useful for predicting the probability of detecting faulty classes
9	Basili et al	1996	Slightly adjusted some of Chidamber & Kemerer's metrics in order to reflect the specificities of C++ language.
10	Abreu and Melo	1996	Studied the MOOD metrics for correlating with the system reliability and maintainability.
11	Harrison and Samaraweera	1996	Defined the usefulness of a design metric (the number of test cases) to predict the testing time.
12	Binkley et al	1998	Applied class coupling method to validate the coupling dependency metrics as a predictor of run time failures and maintainability measures by the usage of C++ system.
13	Tang et al.	1999	Studied the OO metrics by using C & K metrics but not used the LCOM and LOC methods as variable.
14	Muthanna et al.	2000	Investigated the use of software design metrics to statistically estimate the maintainability of large software systems, and to identify error prone modules. The fact that the usage of metrics in the analysis and design of object oriented (OO) software can help designers make better decisions is gaining relevance in software measurement arena.
15	Polo et al	2001	Reported code metrics for prediction of maintenance of legacy programs as a case study by applying logistic regression, MANTEMA a methodology for maintenance developed by atos ODS.

16	Genero et al.	2003	Presented a controlled experiment and its replication, which had carried out to gather the empirical data.
17	Subramanyam and Krishnan	2003	Conducted an empirical analysis on subset of C & K metrics suite in determining software defects.
18	Hayes et al	2004	Proposed a metrics-based software maintenance effort model by the use of validation datasets. They applied COCOMO (constructive cost estimation Model), SLIM, AMEffMo(Adaptive Maintenance Effort Model) and Regression Analysis methods for computing the effort.
19	Kiewkanya et al.	2004	Constructed maintainability model of object-oriented design by applying methods like association, aggregation, generalization and classification.
20	Hayes et al.	2005	Proposed prediction of maintainability using regression analysis and DC ratio methods.
21	Bocco et al.	2005	Assessed the capability of internal metrics as early indicators of maintenance effort through experimentation. The usefulness of measures for the analysis and design of object-oriented software is increasingly being recognized in the field of software engineering research. Recognition of the need for early indicators of external quality attributes is increasing.
22	Genero et al.	2007	Investigated through experimentation whether a collection of UML class diagram measures could be excellent predictors of two main sub distinctiveness of the maintainability of class diagrams namely understandability and modifiability.
23	Breesam	2007	Formalized a set of metrics with the intention of using them for the quality measurement of an object-oriented design with consideration of class inheritance.
24	Neelamegan and Punithavalli	2009	They surveyed four object-oriented quality metrics and found that these metrics focuses on measurements when apply them to the class and design characteristics.

25	Sastry and Saradhi	2010	Attempted to implement software metrics with assistance of GUI and also examined relationships of metrics for determination of quality. The quantity of software attributes estimated by observing the life cycle of object-oriented software development.
26	Amjan Shaik et al.	2010	Statistically analyzed the object-oriented software metrics on CK metric collections by verifying the data put together from the projects of some students. Fast feedback for software designers and managers were provided by applying metrics data. They examined minutely that proper use of these metrics directly reduces the cost of the overall execution and enhancement of quality of the ultimate product.
27	Rizvi S.W.A. et al.	2010	Proposed MEMOOD model that is an improvement of the maintainability or understandability of class diagram and accordingly the maintainability in concluding software.
28	Thapaliyal et al	2010	Analyzed the software defects and object oriented metrics. They evaluated two metrics weighted method per class (WMC) and coupling between object classes (CBO) of C and K metrics Suite. They investigated whether the metrics taken are linked to defects or not by the usage of 50 samples of Java classes of different projects.
29	Gautam C. et al.	2011	Described that the compound MEMOOD model is better than the MEMOOD model to determine the maintainability of class diagram in terms of their understandability, modifiability, scalability and level of complexity.
30	Malhotra R. and Jain	2011	Reviewed software fault prediction for object-oriented system which uses logistic regression method.
31	Shaik et al	2011	Suggested a metric approach for evaluating the system test cases in OO system through carving and replaying the degree of differences in unit test cases. They have used regression CR tool architecture method for the evaluation process.

2.3 LITERATURE SURVEY OF MAINTAINABILITY AND ARTIFICIAL NEURAL NETWORK

Since the preceding years a lot of object-oriented metrics have been explored. These metrics can be employed to obtain assurances about software quality by the usage of prediction models. Maintainability is characteristically measured as change effort which can either be the average effort for making a change to a class, or the overall effort used on changing a class.

In the literature, a number of metrics suggested to emphasize on the quality of object-oriented (OO) design by various researchers. To capture the distinctive facet of OO design, various metrics designed which estimate the quality of the software. These metrics are very fast in assessing the bulky software with little maintenance cost and can be used in earlier phases of software development. However it is necessary to know the metrics that are valuable in capturing significant quality attributes, e.g. effort, productivity, fault-proneness and evaluation of maintainability. The relation between maintainability and metrics can be discovered by using two techniques; either objective or subjective. The metrics theoretically and empirically validated according to the previous studies which evaluates the impact of the metrics on maintainability. Then after predictive models are constructed using ANN like soft computing techniques to estimate the quality attributes of the software system.

OO methodology has been enormously utilized in software engineering. As its use increases rapidly, the current applications need to be enhanced for the further development process. Li and Henry [44] have constructed a prediction model integrating ten object oriented metrics with the help of the statistical analyses technique. The result of their study has established a strong relationship between metrics and maintenance effort in OO systems. Basili *et al.* [45] obtained that the Chidamber and Kamerer metrics were associated with fault proneness. It was based on a study of eight medium-sized systems devised by students. Khoshgaftaar *et al.* [69] introduced application of neural network tool to predict software quality. A huge telecommunications system has been presented by them which classify the modules as either fault prone or not fault prone. They had made an assessment among ANN and a non-parametric discriminate model which revealed that ANN model had better predictive accuracy than the other one. Fenton and Neil [70] evaluated various sizes and complexity metrics related to software fault prediction models to assess the defects. They made comparison of fault-proneness estimation models and concluded that software quality is a vital prerequisite in the system development.

Giovanni [71] correlated a collection of software fault-proneness and static metrics. Statics metrics computed statically in the source code (e.g. Mccabe's cyclomatic complexity) whereas dynamic metrics measure thoroughness of testing (e.g. structural and dataflow coverage). Such metrics partially influences the software fault- proneness and provides restricted support for adjusting the process of testing. Khoshgoftaar *et al.* [72] demonstrated a case study of real time avionics software that could predict the testability of each module by doing static measurements of source code. They observed that ANN is an ensuring technique for constructing predictive models, as they are capable of modelling nonlinear relationships. Emam *et al.* [73] have constructed a model to predict which classes in a future release of a commercial Java application will be faulty. The model was then validated on a subsequent release of the same application. Their results indicated that the prediction model had a high accuracy. Fioravanti and Nesi [74] have extracted over 200 different OO metrics for identifying a suitable model for detecting fault-proneness of classes. They concluded that only little of them were appropriate for identifying fault-prone classes. Genero *et al.* [75] have presented a set of metrics to measure the UML class diagrams' structural complexity and used them for predicting maintainability. This will be correlated with object oriented information system maintainability to a great extent. Briand *et al.* [76] investigated the relations present between OO metrics and the probability of fault detection in system classes empirically. They studied a set of OO metrics in terms of their effectiveness in predicting fault-proneness and empirically validated these metrics as an important software quality indicator. Validation is carried out by them using two kinds of data analysis techniques such as regression analysis and discriminate analysis. They researched about the existing relations of OO with coupling, cohesion, inheritance measures and the probability of fault detection in system classes during testing through empirical observations. Their univariate analysis has shown that many coupling and inheritance measures are strongly related to the probability detecting fault in a class. Their multivariate analysis results showed that by using some of the coupling and inheritance measures, very accurate models could be derived to predict in which classes most of the faults actually lie.

Quah and Thwin [77] predicted the total faults in a particular class by using a multiple regression model and a neural network model. They have used three industrial real-time subsystems data and found that neural network model can predict more accurately than regression model. Yu et al. [78] choose eight metrics to identify their relationship with each other and also determined fault proneness of the software. The subject system was the client side of a large network service management system developed by three professional software engineers. This application was written in Java comprising of 123 classes and about 34,000 lines of code. First of all, they tested the correlation between the metrics and got four extremely correlated subsets. Subsequently, univariate analysis was employed to identify the metric that can perceive the faults. Menzies et al. [79] have compared various methods such as naive Bayes, decision trees, and 1-rule classifier by using the NASA software defect data. Mahaweerawat [80] described fault prediction model based upon supervised learning using multilayer perceptron network and the results are analysed in terms of classification. Faulty classes are again analysed and categorized according to their particular fault.

Gyimothy et al. [81] has done empirical validation of metrics developed by Chidamber and Kamerer [41-43] based upon open-source software used to predict the fault. Both types of regression, i.e. linear type and logistic type regression; and machine learning methods i.e. neural network and decision tree were employed by them to construct models. Bellini [82] compared fault-proneness estimation models by applying the logistic regression and the discriminant analyses methodologies. They have used various datasets according to the requirement. Tian and Noore [83] applied evolutionary neural network modelling to predict collective failure time of software. Thwin and Quah [84] applied neural networks techniques for predicting software quality. They have taken OO metrics as independent variable and used two neural network models, namely ward neural network and General Regression Neural Network (GRNN). They concluded that GRNN network model can predict more accurately than Ward network model.

Yan Ma [85] explored exact estimation of fault prone modules in the software improvement process which enabled effective invention and recognition of defects. Sandhu et al. [86] compared all the classes of WEKA's machine learning algorithms and found that Logistic Model Trees algorithm is the most excellent prediction technique among other classes of machine learning algorithms in prediction of severity of faults in software systems. Hu and Zhong [87] proposed a model to predict menace associated with software module with the help of neural network. Learning vector quantization network was used for the prediction of the software quality. The study by Zhao and Zhong et al [88] suggested a software fault-proneness prediction model by support vector machine and the Chidamber-Kemerer (C&K) object-oriented metrics. Yogesh et al. [89] predicted the testing effort of software quality attribute by applying ANN method and taking CK's OO metrics as the independent variable. They used the public domain data of NASA and estimated the testing effort. Sandhu et al. [90] had explored the impact of faults in OO software modules. The primary objectives of their work are discovering the structural code and design attributes of software systems and getting the best algorithms which can be employed to model the impact of faults in OO software. They also predicted the impact of the defect on the overall environment in function-based systems. They observed that neuro-fuzzy based predictor models are the best one. They used a public domain defect dataset of NASA and coded it in C programming language.

Arvindar et al. [91] predicted the software maintenance effort by applying various soft computing approaches. Two commercial software data sets were used, and it was observed that these soft computing techniques are really useful for the construction of accurate models to predict the maintenance effort. They have chosen maintenance effort as the dependent variable and eight OO metrics as the independent variable. Ratra et al. [92] compared early estimation of fault-prone modules in software designs by using neural network and clustering techniques. They measured the performance of the two methods based on their accuracy, the root means square error and mean absolute error values. They opined that the performance of the neural network approach is quite superior to the approach based upon clustering. In this way, various researchers have proposed different sophisticated modelling techniques by using artificial neural network in the object-oriented software paradigm.

2.4 SUMMARY

It is concluded by various researchers that maintainability is a critical concern in the OO design as it uses a large volume of small methods. Therefore, OO paradigm must be chosen to solve the complexity of measuring the maintainability. This survey result is concurrent with the products of different research that are done to overcome the issues of maintainability estimation in object-oriented systems. The multitude of methods of maintainability summarized in this survey will be a key guide to develop an efficient and cost-effective way of maintaining the object-oriented systems, leading to the betterment of software quality.

Researchers acknowledged the important function of quality estimation in the design of object-oriented software systems. To get over the issues of predicting maintainability in an OO system, suitable and efficient metrics must be defined and chosen. Use of artificial neural networks will be an efficient method to estimate maintainability in an object-oriented system. Among the different soft computing techniques, ANN possesses advantages of predicting the software maintenance effort by minimal computation. It can be applied as a predictive model due to its incredible representation techniques and capability to execute complicated functions. In addition, it is among one of the emerging methods in software development and demonstrates a crucial role in predicting software quality.

Chapter 3

DESIGN AND IMPLEMENTATION

3.1 INTRODUCTION

Statistical modeling techniques have a vital role in clear understanding of software development and design. Among them neural network techniques have enhanced predictive capability than most other statistical models. As object-oriented systems use many small methods, a unique maintenance difficulty is associated with it. The association between OO metrics and software maintenance effort is complex and nonlinear [93]. Hence this study is conducted on OO paradigm, which is a good platform to assess maintenance effort. As we know, measuring the object-oriented software is very crucial for the software engineers. The necessity of measurement is to reduce cost, effort and schedule for amplifying the system performance. Maintainability of software can be obtained upto some extent by the use of available commercial techniques. There are various software metrics in the object-oriented system which can measure maintainability. The prediction of maintainability requires analysis of huge volume of data; hence this is time consuming and faulty. The main goal of PCA is to reduce that raw data by removing the correlations among the software metrics [69], thereby making the prediction model more stable.

In fact, quality estimation signifies either measuring reliability or maintainability of the software. Maintainability is usually measured as the change in effort i.e. the number of lines altered per class. Artificial Neural network (ANN) is used as a predictive model because of its minimal computation and complex modeling ability. Therefore, we established an artificial neural network model to improve prediction outcome of estimating software maintainability.

3.1.1 GOAL OF THE STUDY

The goal of the present study is to assess the maintainability of object-oriented metrics by using ANN technique.

3.1.2 OBJECTIVES

Objective 1: To familiarize yourself with variety of object-oriented metrics, their definitions and operations.

Objective 2: To consider earlier studies for validating the metrics theoretically and empirically.

Objective 3: To design a methodology to evaluate the metrics i.e. adoption of Artificial Neural Network (ANN) technique for measuring maintainability of object-oriented system.

Objective 4: To compare the result and efficacy of the proposed model with the previously existing models.

3.2 RESEARCH METHODOLOGY

The proposed research is carried out by initial entry of raw metric data. Then all erroneous data, missing values and redundancy from those data were removed by making them normalized. The principal components analysis (PCA), a data reduction technique was used with the intention of reducing the dimensionality problem of multivariate data. This method not only reduces the complexity of data but also reduces the variables used in the study. The analysis identifies the latent factors from a group of apparent important variables. In an extensive way it is a technique that analyses the correlation between variables and produces a few factors of the original data. Hence it can be opined that this method is more economical in the sense of storing the data and saving memory space. Accurate neural training can be done by transferring the raw data into principal components. The significance of principal components analysis was illustrated with the help of two commercial raw datasets, namely UIMS (User Interface Management system) and QUES (Quality Evaluation Systems) [44]. Subsequently, neural network modeling was described. The maintenance effort data was collected from a commercial system, which is designed and implemented using Classic-Ada. Classic-Ada is an object-oriented programming language that provides object-oriented constructs as well as Ada constructs.

The UIMS (User Interface Management System) data was used for this study. The maintenance effort was measured by the number of lines changed per class. This measurement was used to estimate maintainability of the object-oriented system. The UIMS system comprises 39 classes and the QUES system consists of 71 classes. After performing principal component analysis, the obtained domain metrics were fed into the neural network as input data. The ANN was trained by applying operative modalities on the domain OO metric data. Finally, we can assess the maintainability of object-oriented software systems. Figure 3.2 explains the flow diagram of steps being followed.

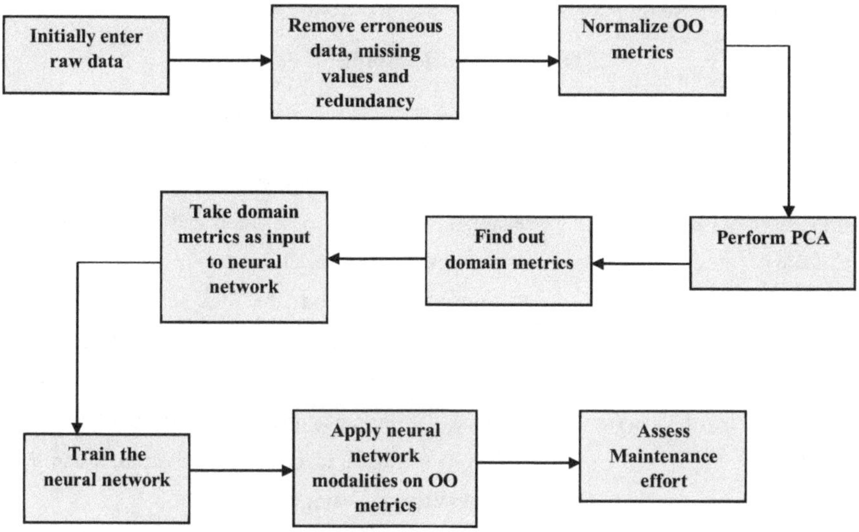

Figure 3.2: Process flow diagram of steps followed

3.3 OBJECT ORIENTED METRICS CHOSEN FOR ANALYSIS

Quality estimation means estimating maintainability or reliability of software [3]. Several researchers have aimed to specify maintainability by using various kinds of measures. Chidamber & Kemerer (C&K) [41-43] have explored some of the object-oriented design metrics for computing maintainability effort. These metrics include WMC, DIT, NOC, CBO, RFC and LCOM. The four additional metrics such as MPC, DAC, NOM, and SIZE 2 were proposed by Li and Henry [44]. SIZE1 is a traditional line of code size metrics. The maintainability was measured in terms of CHANGE metrics, which inform the number of lines changed per class. The definition of each metric used in the dataset is described in Table 3.3. Metrics proposed by Chidamber & Kemerer [41-43] and Li and Henry [44] were considered in this investigation. Size 1, the traditional metrics were also chosen for the study.

Table 3.3: Metrics Studied

Metric	Definition
DIT (Depth of Inheritance Tree)	The maximum depth of the inheritance hierarchy for each class.
NOC (Number of Children)	The number of direct subclasses (children) of a class.
MPC (Message Passing Coupling)	Measures the number of method calls defined in a class that invoke methods of other classes.
RFC (Response for a Class)	The number of methods that can potentially be executed in response to a message received by an object of the class, including methods in the class itself and those it calls.
LCOM (Lack of Cohesion in Methods)	Measures the lack of cohesion among methods in a class by evaluating the extent to which methods share instance variables. A higher value indicates low cohesion.
DAC (Data Abstraction Coupling)	Measures the number of abstract data types (ADTs) used by a class.
WMC (Weighted Methods per Class)	The sum of the complexities of all methods in a class, where complexity is often measured using cyclomatic complexity.
NOM (Number of Methods)	The total number of methods defined in a class, including inherited and locally declared methods.
SIZE1	The number of attributes (instance variables) in a class.

3.4 PRINCIPAL COMPONENTS ANALYSIS

This section explains the application of principal component analysis to neural network modeling as a way to improve the predictability of neural networks. The purpose of principal component analysis is to augment the performance of discriminant models of software quality.

The data of a software system is highly interlinked. If there is a high correlation between the data sets, this indicates that these variables are aiming to describe the same goal and class property of the object to be measured. Similar is the situation with object-oriented systems, where many metrics used to predict maintainability are strongly correlated with each other. Suppose there are **n** classes and **m** number of metrics, then the total data set will be **n x m**. This will be a huge data set to analyze.

Again, the presence of several correlated data is merely duplicating the complexity for analysis. Thus, there is a need for reduction of the data. To focus or analyze the minimum data, there is a need for an effective tool to sort out the principal component, which will represent the original **n x m** data. Here comes the role of PCA i.e. principal component analysis.

PC analysis can be understood as:

1) It converts raw metrics to variables that are not linked to each other. So, the PC data is uncorrelated or orthogonal.

2) It is a data reduction technique. It converts **n x m** data set to **n x p** data set, where **p<m**. This is by pulling out linear combinations of the original data in such a way that they account for most of the original disparity in the dataset.

Thus, Principal component analysis (PCA) is a standard technique to identify the underlying, independent/orthogonal dimensions that explain relations between the variables in a data set [84].

PC analysis is applied to maximize the sum of squared loadings of each factor extracted in turn. In PC analysis, a new variable (Pi) is constructed, called Principal

Component (PC) from a given set of variables Xj' s(j = 1,2,....,k) as shown in equation 3.4.1.

$$\mathbf{P_1 = b_{11} X_1 + b_{12} X_2 + \ldots\ldots + b_{1K} X_K}$$
$$\mathbf{P_2 = b_{21} X_1 + b_{22} X_2 + \ldots\ldots + b_{2K} X_K}$$

$$. \quad .. \quad .. \quad .. \quad .. \quad .$$

$$\mathbf{P_k = b_{k1} X_1 + b_{k2} X_2 + \ldots\ldots + b_{KK} X_K} \qquad \mathbf{(3.4.1)}$$

The factor b_{ij}s' are called loadings and are found out in such a manner that the extracted PCs must be uncorrelated.

The variables associated with high loadings require some degree of interpretation. To identify these loaded variables and to interpret the PC's, rotated components are considered. As the PCs are independent of each other, orthogonal rotation is applied. There are various strategies to perform such rotation and according to literature, varimax rotation is the most frequently used strategy. Eigenvalue or latent root is an integral part of PC. The summation of squared values of loadings describing the dimension is referred to as the eigenvalue. PCs with eigenvalue greater than 1 are usually considered for interpretation [94].

Given that **R** is a real symmetric matrix and that it has distinct roots, it can be stated in equation 3.4.2.

$$\mathbf{R = P.\Lambda.P^T} \qquad \mathbf{(3.4.2)}$$

In this, the eigenvalues $\boldsymbol{\lambda}$ lie on the diagonal of $\boldsymbol{\Lambda}$. PC_j is the eigenvector linked with λ_j. Orthogonal linear combinations of the original data are presented by the coefficients of **m** eigenvectors in PC. These linear combinations are the PC of **R**. Fraction of variance in the software measures is $\boldsymbol{\lambda_j/m}$, which is explained by PC_j. The diagonal of Λ forms a decreasing list that explains all of the data variance as shown in equation 3.4.3,

$$\boldsymbol{\lambda_1 > \lambda_2 > \ldots > \lambda_m}. \qquad \mathbf{(3.4.3)}$$

The total variance is the sum of the eigenvalues as observed in equation 3.4.4,

$$\sum_{i=1}^{m} \lambda_j = m \ [1] \qquad (3.4.4)$$

A large proportion of the total variance is usually explained by the first few PC. So, to reduce the dimensionality of the problem without a substantial loss of the explained variance, restriction of analysis can be done to a relatively few PC.

That is why most often a common stopping criterion is followed, i.e. to consider PC with eigenvalues greater than one. In another way, the number of PC taken can be determined by the cumulative amount of variance associated with the PC [72]. For example, PCs that account for greater than 70% of the explained variance are included for interpretation.

$$PC = z.T \qquad (3.4.5)$$

In PC analysis **T** (in the above equation 3.4.5) describes the relationship between each of the **m** variables and the p significant PC. If **z** is the matrix of m standardized observed software measures, then this **z** can be reduced to **p** orthogonal PCs.

Most of the variations in the raw metric data are accounted for by these components. Thus, the software measures (metrics) are transformed in this manner to component/domain measures (metrics), and these domain measures are to be used as the input variables in the neural network.

3.4.1 Application of PCA in Neural Network Modelling

Various measures have been proposed by researchers to capture the quality of object-oriented design and code [96-103]. The explored methods used to identify fault-proneness of the system by the aid of artificial neural networks. Several analyses have also been done to predict software quality by using statistical methods. There is an enormous increase in the research about ANN over the years and successfully applied across a series of problem domains such as finance, medicine, engineering, geology, and physics. Indeed, for the problems of prediction, classification and control, neural networks will be an efficient asset to solve them [95]. Some of the applications of PCA to neural network modelling are listed in Table 3.4.1.

Table 3.4.1: Studies on PCA and Neural Network

Sr. no.	Author	Reference	Year	Description
1	Khoshgaftaar *et al.*	[96]	1994	They explored the potential of PCA in improving Neural Network Predictions of Software Quality. It was established that PCA resulted in quicker training of the neural network.
2	Lanubile *et al.*	[101]	1995	They compared models to categorize fault-prone software components. They applied various modeling techniques in their study including logistic regression, discriminant analysis, layered neural networks, principal component analysis, holographic network, and logical classification models. They found that no model can separate components from faults and without faults.
3	Khoshgaftaar *et al.*	[69]	1997	Neural network was applied by them for software quality modeling of a very huge telecommunications system. They have used PCA to remove the high correlation of the software metrics.
4	Khoshgaftaar and Szabo	[72]	2000	They have used neural networks for predicting software faults during testing. Ten software product measures were compared using two neural networks, i.e. raw analysis data and principal component analysis. It was found that PCA has better predictive quality.
5	Briand *et al.*	[102]	2001	They investigated quality factors in object-oriented designs by applying various techniques such as principal component analysis, logistic regression, and univariate regression analysis on an open multi-agent development environment: LALO (language agents Logical Object) which consists of 90 classes and 40K SLOC.
6	Quah *et al.*	[84]	2003	Neural Networks models were used for the prediction of software quality using object-oriented metrics. They have used two neural networks (WARD and GRNN) and proposed that GRNN has better predictability than the WARD network. In this study, they have used PCA with varimax rotation method for QUES

				(Quality Evaluation System) and HMI (Human Machine Interface) systems.
7	Thwin and Quah	[97]	2005	They have applied neural networks for maintainability prediction of software and also used PCA. They have taken OO metrics as the independent variable and used two models of neural network, namely Ward neural network, and general regression neural network (GRNN). They concluded that the GRNN model can expect more accurately than the Ward network model. The data systems used are QUES and UIMS (User Interface Systems).
8	Aggarwal *et al.*	[94]	2006	They have applied artificial neural network (ANN) to predict maintainability using object-oriented metrics. In their study, maintenance effort was used as a dependent variable and the principal component of 8 object-oriented metrics as the independent variable. They establish that ANN is very useful in software quality prediction.
9	Aggarwal *et al.*	[98]	2007	They investigated the effect of design metrics on fault proneness in object-oriented systems. They have collected data from Java applications which contained 136 classes. They have used PCA and extracted up to 6 principal components. Their proposed model predicted faulty classes with more than 80% accuracy.
10	Katiyar *et al.*	[103]	2010	They estimated software quality by applying object-oriented metrics through an artificial neural network. They have used the principal mechanism of eight OO metrics as the independent variables and established that ANN method was valuable for building a software quality model.
11	Shaik *et al.*	[99]	2011	They investigated the result of object-oriented design software metrics on fault proneness in object-oriented systems. They have applied the principal component method to object-oriented metrics to get the fault proneness. They have extracted up to 6 components and found that the number of dimensions capture

				in PCA is much less than the amount of metrics.
12	Zhong *et al.*	[100]	2011	They tried to predict software quality by applying a major components study, a wavelet neural network, and a genetic algorithm. They stated that the model combining principal components analysis with wavelet neural network and genetic algorithm can obtain better prediction accuracy than the models using neural network of back propagation and generalized regression neural network.

3.5 NEURAL NETWORK MODELLING

The Neural network is a network of interconnected neurons where information propagation occurs by firing electrical pulses via its connections. During the lifetime of a neuron, the connections or weights need to be adjusted [95]. The neural net consists of one input layer for feeding raw data to the network, one hidden layer for computing that data and one output layer for getting the computed results. The neural network is distinguished by the weighted connections among neurons. [29]

If the connection between **i** and **j** neuron is **Wij**, then input value at neuron **i** will be the weighted sum of the entire inputs as stated in equation 3.5.1, where θ_i is the threshold of neuron **i** [96].

$$\mathbf{Net_i = \sum_j W_{ij} * Out_j - \theta_i} \qquad (3.5.1)$$

The output of neuron **i** is the sigmoid function where **T** adjusts the temperature or gain of the function as shown in equation 3.5.2.

$$\mathbf{Out_i = \frac{1}{(1+e^{-Net_i}/T)}} \qquad (3.5.2)$$

The error in the neural network can be defined by using the formula stated in equation 3.5.3.

$$\frac{1}{2}\sum_{i=1}^{m} \mathbf{(d_i - Out_i)^2} \qquad (3.5.3)$$

Where **M** = Number of output neurons,

$\mathbf{d_i}$ = Desired output for the ith output of neuron

$\mathbf{Out_i}$ =Actual output.

The ANN is employed for prediction of maintenance effort. All the domain metrics are used for input to the neural network. If the correlated raw object-oriented software metrics are used directly, then the neural net models will not train properly [69]. The correct neural training can be done by transferring the raw data into principal components.

3.5.1 Multi-Layer Perceptron (MLP) Network Model

The Neural network is a network of interconnected neurons where information propagation occurs by firing electrical pulses via its connections. During the lifetime of a neuron, the connections or weights need to be adjusted. The neural net consists of one input layer to feed raw data to the network, one hidden layer to compute that data, and one output layer for getting the computed results. The neural network is characterized by the weighted connections between neurons [95]. The Multi-Layer Perceptron network is a function of independent variables which minimize the prediction error of output variables. This network is also called feed forward architecture as it does not contain any feedback loops. The connections flow forward from the input layer towards the output layer in the system.

Generally, the ANN adjusts the weights values to produce a specific target output from a particular input [95]. Standard back-propagation algorithm with momentum of 0.9 is used to train the MLP model. The minimum square error (MSE) is followed as the training stopping criterion.

3.5.2 Learning Algorithm Used

During the lifetime of a neuron, the connections or weights need to be adjusted. Similarly, the quantity of incoming pulses required in activating a neuron is changed. This is the behavioural attributes that let the NN to learn. There are three basic learning algorithms in neural networks, namely supervised learning, unsupervised learning, and reinforcement learning. Supervised learning is most commonly used, and also called as learning with a teacher. It is applied when the target value is known and associated with each input in the training set.

3.5.2.1 Error Back-Propagation Algorithm

The error back-propagation algorithm is a well-known method for training hidden units of a Multi-Layer Perceptron network. It is a novel indication in artificial neural network research and its applications [104].

If we perform training operation on a multi-layer feed forward network or multi-layer perceptron, then the gradient descent method is applied to approximate the unknown function. This is based on several training datasets which comprise of pairs (**x**,**t**). A pattern of input to the neural network is represented by vector **x** and consequently the target output is denoted by vector **t**. The total gradient that concerns the whole training set is the summation of the gradients for every pattern. The weight from unit j to unit i is denoted by w_{ij}. The detailed steps followed are described below [104].

STEPS FOLLOWED:

1. **Data structures used:**

Data structures used

1	The error signal for unit j	$\frac{\partial E}{\partial net_j}$
2	The gradient for ij	$-\frac{\partial E}{\partial w_{ij}}$
3	The set of nodes anterior to unit	$(j : \exists w_{ij})$
4	The set of nodes posterior to unit	$(i : \exists w_{ij})$

2. **The gradient:**

To train the neural network, the gradient G of the loss function relating to each weight w_{ij} of the network should be calculated. It describes how a minute alteration in those weights will influence the total error (E). The gradient is expanded in 2 factors by applying the chain rule as given in equation 3.5.2.1.

$$\Delta w_{ij} = -\frac{\partial E}{\partial net_j}\frac{\partial net_j}{\partial w_{ij}} \qquad \textbf{(3.5.2.1)}$$

The initial factor is the error of unit i. The next factor is given in equation 3.5.2.2

$$\frac{\partial net_j}{\partial w_{ij}} = \frac{\partial}{\partial w_{ij}}\sum_{k\in Ai} w_{ik}y_k = y_j \qquad \textbf{(3.5.2.2)}$$

Putting the two equations in combination, we obtain equation 3.5.2.3

$$\Delta w_{ij} = \delta_i y_j \qquad \textbf{(3.5.2.3)}$$

For computing this gradient, the activity as well as the error for all significant nodes in the network is to be known.

3. Forward activation: The activity of the input units is decided by the external input **x**. The activity is propagated forward for the entire added units. This is shown in equation 3.5.2.4.

$$y_i = f_i\left(\sum_{j\in Ai} w_{ij}y_j\right) \qquad \textbf{(3.5.2.4)}$$

The activity of each anterior node (which forms the set A_i) must be known before the activity of unit i can be worked out. There is an ordering of nodes in feed forward networks from input to output as they do not have cycles.

4. Calculation of output error: If it is assumed that the sum-squared loss is used, then the error is calculated as in a way as given in equation 3.5.2.5.

$$E = \frac{1}{2}\sum_{0}(t_o - y_o)^2 \qquad \textbf{(3.5.2.5)}$$

The error for output unit o is just like that of equation 3.5.2.6.

$$\delta_o = t_o - y_o \qquad \textbf{(3.5.2.6)}$$

5. Error back-propagation: The error must be propagated back from output nodes for hidden units. Therefore, this algorithm is named back-propagation algorithm. Again, applying the chain rule, the error of a hidden unit can be expanded in form of its posterior nodes as given in equation 3.5.3.7.

$$\delta_j = -\sum_{i \in P_j} \frac{\partial E}{\partial net_i} \frac{\partial net_i}{\partial y_j} \frac{\partial y_j}{\partial net_j} \qquad \textbf{(3.5.2.7)}$$

Among the 3 factors inside the summation, the first factor is simply the error of node i. The second factor is presented in equation 3.5.2.8.

$$\frac{\partial net_i}{\partial y_j} = \frac{\partial}{\partial y_j} \sum_{k \in Ai}^{n} w_{ik} y_k = w_{ij} \qquad \textbf{(3.5.2.8)}$$

However, the third one is the derivative of node j's activation function, which is given in equation 3.5.2.9

$$\frac{\partial y_j}{\partial net_j} = \frac{\partial f_j(net_j)}{\partial net_j} = f_j'(net_j) \qquad \textbf{(3.5.2.9)}$$

For hidden units h which uses the hyperbolic tangent (tanh) as activation function, the special identity, $\tanh(u)' = 1 - \tanh(u)^2$ can be used to give equation 3.5.2.10.

$$f_h'(net_h) = 1 - y_h^2 \qquad \textbf{(3.5.2.10)}$$

Putting all the parts together we obtain equation 3.5.2.11.

$$\delta_j = f_j'(net_j) \sum_{i \in Pj} \delta_i w_{ij} \qquad \textbf{(3.5.2.11)}$$

To compute the error for unit j, primarily the error of all its posterior nodes must be known, which forms the set, P_j.

3.6 PREDICTION OF MAINTENANCE EFFORT BY MLP MODEL

Maintenance effort of two commercial software products such as UIMS and QUES [44] was estimated in this study. The maintenance effort is estimated by considering the number of lines altered per class. A change in a line can be an addition or a deletion in the class. This dimension is used in the study to evaluate the maintainability of the object-oriented systems. Data sets were divided into training, testing, and production sets using a 7:2:1 ratio. The production data set is used to evaluate model performance. We develop a MLP network model and compared our model results with Ward and GRNN neural network results.

3.7 SUMMARY

Improvement in the software development and the software quality needs a higher range of precision during prediction. Several statistical models, including artificial neural networks, are now becoming a part of software engineering for prediction and measurement. This process of estimation is being overwhelmed with a large number of software metrics, a huge number of executable statements, and lines of code. Furthermore, these data often tend to have a higher degree of correlations among themselves, which makes the prediction process more complex. To get around these problems, principal component analysis has shown a way to most of the researchers. On one hand, it reduces the multidimensional problem of the huge data by converting it to a small orthogonal dataset, on the other hand, it finds out the factorial relationship and most of the variance of the original data set. The resultant orthogonal domain metrics provide a swift base for the easier prediction by neural networks. Thus, principal component analysis has incredible importance in evaluating the original data and predicting the software quality. Similarly, the back-propagation algorithm also has its significance, which is applied to train the neural network model.

Chapter 4

EXPERIMENTATION

4.1 INTRODUCTION

This section briefly describes the software used for this investigation and results in the form of snapshots. SPSS software is used for analysis purposes. This parcel of programs is available for personal and mainframe computers. It is not only a comprehensive system that analyzes data but also comprises a series of software tools for managing the data, analyzing those data statistically. Both datasets UIMS and QUES were used in this study to assess the maintainability of object-oriented software. The snapshots of the results obtained by using the SPSS software are also presented in this chapter. The results obtained by Principal Component Analysis (PCA) and the Multi-layer perceptron (MLP) neural network modeling were presented, and various validation tests were performed to measure maintainability of the object-oriented software.

4.2 SPSS SOFTWARE

SPSS is the Statistical Package for Social Science. It was developed by H. Nie and C. Hadlai Hull in 1968. Presently IBM SPSS is completely integrated into the IBM Corporation. It is one of the brands which are under IBM Software Group's Business Analytics Portfolio. It is a wide range and supports flexibility during analysis. It is predictive software which can receive data from any kind of folder and utilize it to produce descriptive statistics and accomplish complex statistical analyses including neural networks. This software is accessible from various platforms such as UNIX, Macintosh, and Windows systems [105]. The details regarding this software are presented in Table 4.4.

Characteristics of SPSS

1) Ease of learning and flexibility.
2) Comprises a series of data managing and editing tools.
3) It provides in-depth statistical capabilities.
4) Presents entire features including plotting, reporting, and presentation.

5) Well, known for its predictive analysis technique.

Table 4.2: SPSS Software Details

Details of SPSS Software	
Developer	IBM Corporation
Initial release	1968
Stable release	20.0 August 16, 2011 (2011-08-16)
Operating system	Windows, Linux, Linux/ UNIX & Mac
Platform	Java
Analysis types	Statistical analysis, Data Mining, Text Analytics, Data Collection, Collaboration & Deployment, and Predictive analysis.
License	Proprietary software
Website	www-01.ibm.com/software/analytics/spss

4.2.1 Data Editor Window

The data editor window demonstrates the contents of a data file. A new data file can be created, or the existing data files can be modified with the assistance of this data editor window. The GUI (graphical user interface) of SPSS includes two kinds of views, such as data view and variable view [105-106].

Data View

The Data View illustrates the rows (cases) and columns (variables) in a spreadsheet view. Contrasting to spreadsheets, here the data cells can only have facts and formulas that cannot be accumulated into these cells.

Variable View

The Variable View demonstrates the metadata glossary where every row not only represents a variable but also displays the name of the variable, label of variable, type of measurement, and a range of additional features.

Both views can be edited manually only by defining the file structure and then allowing data entry without using command syntax. In Figure 4.2.1, the snapshot of the data editor window is presented.

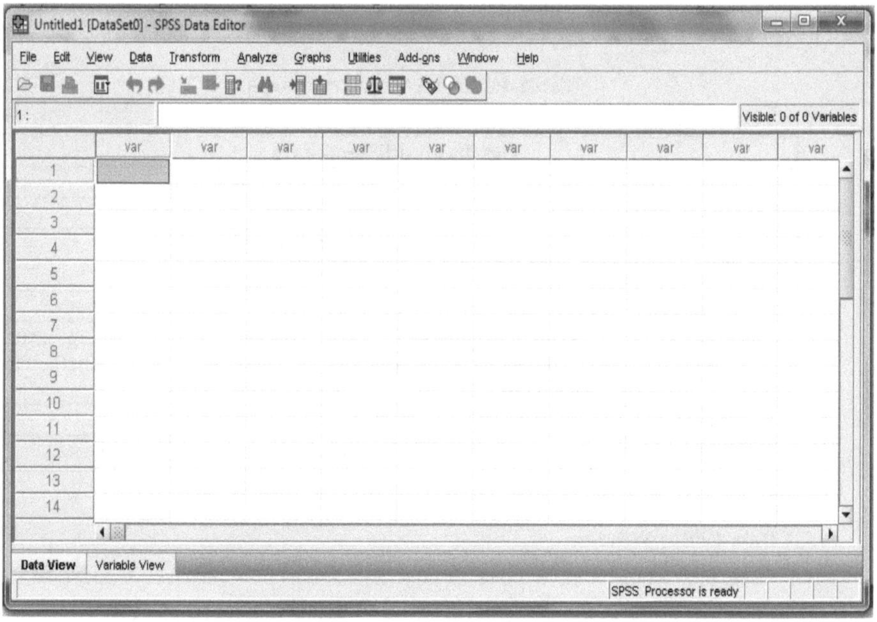

Figure 4.2.1: SPSS Data Editor Window

Syntax editor, an exceptional facet of SPSS, is applied for editing the command syntax. These commands are also stored in a file for further reference. Results are displayed in SPSS by using its output viewer.

4.3 SNAPSHOTS OF THE STUDY

This section presents all the snapshots of the results obtained by applying the SPSS software.

4.3.1 UIMS System

All the snapshots that are taken for the UIMS dataset are given under this section. The detailed neural network information, the MLP network architecture, normalized importance of variables, independent variable importance and correlations between change effort and predicted value are represented in Figure 4.3.1.1, Figure 4.3.1.2, Figure 4.3.1.3, Figure 4.3.1.4, and Figure 4.3.1.5 respectively.

Network Information

Input Layer	Covariates	1	REGR factor score 1 for analysis 1
		2	REGR factor score 2 for analysis 1
		3	REGR factor score 3 for analysis 1
	Number of Units[a]		3
	Rescaling Method for Covariates		Adjusted Normalized
Hidden Layer(s)	Number of Hidden Layers		2
	Number of Units in Hidden Layer 1[a]		2
	Number of Units in Hidden Layer 2[a]		2
	Activation Function		Hyperbolic tangent
Output Layer	Dependent Variables	1	change
	Number of Units		1
	Rescaling Method for Scale Dependents		Normalized
	Activation Function		Sigmoid
	Error Function		Sum of Squares

a. Excluding the bias unit

Figure 4.3.1.1: Neural Network Information (UIMS)

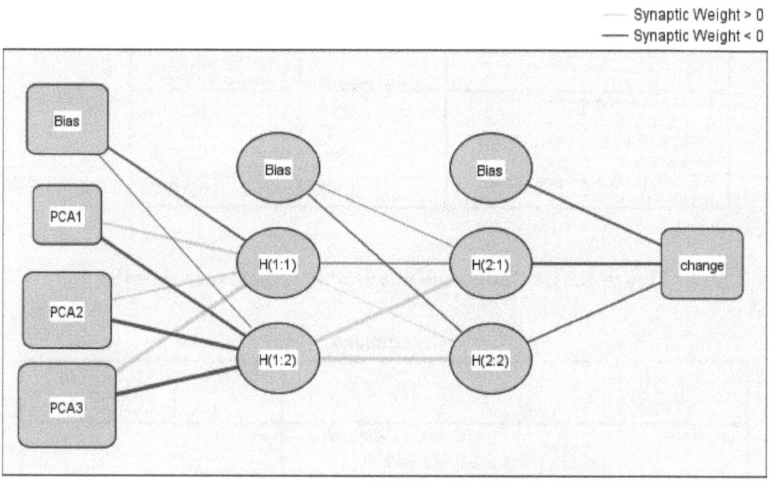

Hidden layer activation function: Hyperbolic tangent

Figure 4.3.1.2: Neural Network Architecture (UIMS)

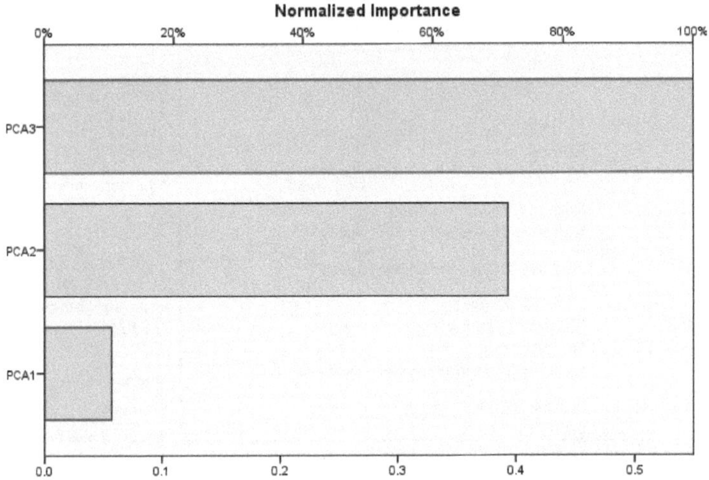

Figure 4.3.1.3: Normalized Importance (UIMS)

Independent Variable Importance

	Importance	Normalized Importance
REGR factor score 1 for analysis 1	.057	10.4%
REGR factor score 2 for analysis 1	.393	71.5%
REGR factor score 3 for analysis 1	.550	100.0%

Figure 4.3.1.4: Independent variable Importance (UIMS)

Correlations

		change	Predicted Value for change
change	Pearson Correlation	1	.946**
	Sig. (2-tailed)		.000
	N	39	39
Predicted Value for change	Pearson Correlation	.946**	1
	Sig. (2-tailed)	.000	
	N	39	39

**. Correlation is significant at the 0.01 level (2-tailed).

Figure 4.3.1.5: Correlations (UIMS)

4.3.2 QUES System

All the snapshots that are taken for QUES dataset are given under this section. The detailed neural network information, the MLP network architecture, normalized importance of variables, independent variable importance and correlations between change effort and predicted value are represented in Figure 4.3.2.1, Figure 4.3.2.2, Figure 4.3.2.3, Figure 4.3.2.4, and Figure 4.3.2.5 respectively.

Network Information

Input Layer	Covariates	1	REGR factor score 1 for analysis 1
		2	REGR factor score 2 for analysis 1
		3	REGR factor score 3 for analysis 1
	Number of Units[a]		3
	Rescaling Method for Covariates		Adjusted Normalized
Hidden Layer(s)	Number of Hidden Layers		2
	Number of Units in Hidden Layer 1[a]		2
	Number of Units in Hidden Layer 2[a]		2
	Activation Function		Hyperbolic tangent
Output Layer	Dependent Variables	1	CHANGE
	Number of Units		1
	Rescaling Method for Scale Dependents		Normalized
	Activation Function		Sigmoid
	Error Function		Sum of Squares

Figure 4.3.2.1: Neural Network Information (QUES)

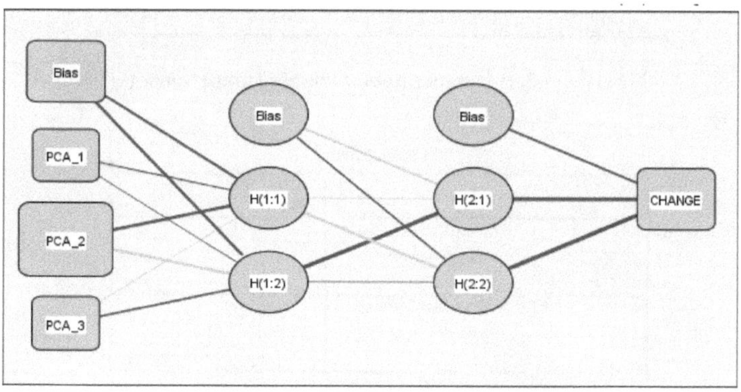

Hidden layer activation function: Hyperbolic tangent

Figure 4.3.2.2: Neural Network Architecture (QUES)

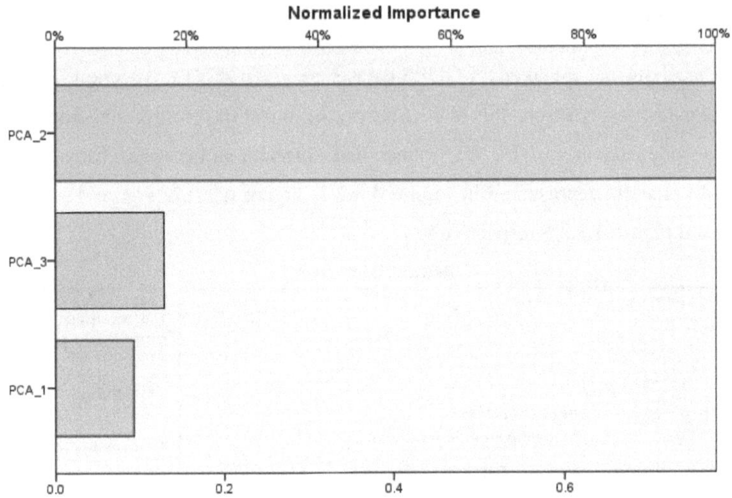

Figure 4.3.2.3: Normalized Importance (QUES)

Independent Variable Importance

	Importance	Normalized Importance
REGR factor score 1 for analysis 1	.075	11.7%
REGR factor score 2 for analysis 1	.639	100.0%
REGR factor score 3 for analysis 1	.287	44.9%

Figure 4.3.2.4: Independent variable Importance (QUES)

Correlations

		CHANGE	MLP_PREDICTED
CHANGE	Pearson Correlation	1	.976**
	Sig. (2-tailed)		.000
	N	71	71
MLP_PREDICTED	Pearson Correlation	.976**	1
	Sig. (2-tailed)	.000	
	N	71	71

**. Correlation is significant at the 0.01 level (2-tailed).

Figure 4.3.2.5: Correlations (QUES)

4.4 EXPERIMENTAL RESULTS OF PRINCIPAL COMPONENTS ANALYSIS

In this study, DIT, NOC, MPC, RFC, LCOM, DAC, WMC, NOM, SIZE1 and SIZE2 are used in UIMS (User Interface System) dataset [44] to produce principal components whereas in QUES (Quality Evaluation system) system all the metrics are studied except NOC. This is because of the relevance of the metrics on the datasets. The rotated component matrix was extracted from all the metrics by using the varimax rotation method. The values above 0.7 are shown in bold in Table 5.2.1, and 5.2.2.

These are the only metrics that are employed to interpret the PCs. For each principal component, we also derived its eigenvalue, percentage of variance and percentage of cumulative variance.

4.4.1 UIMS Result

The interpretations of PCs for UIMS system are described as follows:

1) PC1: RFC, LCOM, WMC, NOM, and SIZE2 are coupling, cohesion, complexity, and size measure metrics. These five metrics were applied for interpreting PC1.

2) PC2: DAC and SIZE1 are the data and size metrics which interpret PC2.

3) PC3: DIT and NOC are inheritance metrics and were used to interpret PC3.

In UIMS system, PCA captures 86.4% of the data set cumulative variance. The detailed results of the principal component analysis are shown in Table 4.4.1.

Table 4.4.1: Principal Components (UIMS)

Rotated Component Matrix			
	Principal Components		
Metrics	**PC1**	**PC2**	**PC3**
DIT	-0.017	-0.349	**-0.790**
NOC	0.125	0.021	**0.852**
MPC	0.683	0.337	-0.257
RFC	**0.876**	0.427	0.084
LCOM	**0.884**	0.082	0.129
DAC	0.271	**0.903**	0.240
WMC	**0.938**	0.175	0.161
NOM	**0.752**	0.580	0.195
SIZE1	0.651	**0.710**	0.217
SIZE2	**0.930**	0.269	0.087
Eigenvalues	6.336	1.558	0.745
% of Variance	63.364	15.584	7.452
Cumulative % of Variance	63.364	78.948	86.400

4.4.2 QUES Result

The interpretations of PCs for QUES system are described as follows:

1) PC1: RFC, LCOM, DAC, WMC, NOM, SIZE1 and SIZE2 are coupling, cohesion, complexity and size measure metrics. These five metrics were applied for interpreting PC1.

2) PC2: MPC is the data and size measure metrics which interpret the PC2.

3) PC3: DIT is the inheritance measure metrics, and was used to interpret the PC3.

In QUES system, PCA captures 86% of the data set cumulative variance. The detailed results of the principal component analysis are shown in Table 4.4.2.

Table 4.4.2: Principal Components (QUES)

Rotated Component Matrix			
	Principal Components		
	1	2	3
DIT	0.097	-0.021	**0.954**
MPC	-0.109	**0.857**	0.151
RFC	**0.844**	0.390	0.049
LCOM	**0.886**	-0.095	0.015
DAC	**0.809**	0.053	0.402
WMC	**0.787**	0.369	-0.270
NOM	**0.984**	-0.046	0.044
SIZE2	**0.975**	-0.004	0.143
SIZE1	**0.746**	0.608	-0.075
CHANGE	0.158	0.826	-0.169
Eigenvalues	5.608	1.872	1.129
% of Variance	56.084	18.719	11.285
Cumulative % of Variance	56.084	74.803	86.088

4.5 ASSESSMENT OF MAINTAINABILITY

The artificial neural network is employed for prediction of maintenance effort. All the domain metrics are used as input to the neural network. If the correlated raw object-oriented software metrics are used directly, then the neural net models will not train properly [69]. Accurate neural training can be done by transferring the raw data into principal components. The maintainability is estimated by using MLP neural network model and the experimental validity is tested for both the datasets in section 5.3.1 and 5.3.2.

4.5.1: Validity Tests Performed

Various validity tests are performed to measure the model performance. Coefficient of multiple determinations (R Square), a statistical indicator used to compare the accuracy of the model by taking the mean of all samples. Table 4.5.1 shows the R square values and their representations [107-108].

Table 4.5.1: R Square Value

R Square Value	
Perfect Fit	1
Very Good Fit	Near 1
Very Poor Fit	Less Than 0

The r- Correlation Coefficient is the most extensively used method, which measures the degree of relationship between variables. Its value ranges from +1 (positive correlation) and -1 (negative correlation). r- Square represents the strength of the relationship and is often called the coefficient of determination. The Mean Square Error is obtained by dividing the sum of the squared errors by the total number of observations. Root Mean Square Error (RMSE) is the square root of the mean square error. The Mean absolute Error (MAE) is a measure used to determine the closeness of predictions to the outcomes. It is given in equation 4.5.1.

$$\mathrm{MAE} = \frac{1}{n}\sum_{i=1}^{n} | f_i - y_i | \qquad \textbf{(4.5.1)}$$

Where f_i is the predicted value and

y_i the actual value

Similarly, the Minimum absolute Error and the Maximum absolute Errors are computed from the results. The p-value is a measure of consistency obtained by calculating the probability of observation of the results of the sample data [107-108]. All these validity tests are performed after getting the predicted value of maintainability by MLP model.

4.5.2 Experimental result for UIMS system

The detailed experimental results obtained by using UIMS dataset are presented in Table 4.5.2. The object-oriented metrics used in this study are DIT, NOC, MPC, RFC, LCOM, DAC, WMC, NOM, SIZE1 and SIZE2. All the validity tests described in section 4.5.1 are carried out to measure the system validity.

Table 4.5.2: UIMS Experimental Result

Validity tests	**MLP MODEL**
R Square	0.8274
r-Correlation Coefficient	0.946
r-Square	0.8949
Mean Square Error	532.565
Root Mean Square Error	23.0773
Mean absolute Error	17.8605
Min absolute Error	0.023
Max absolute Error	67.104
p-Value	0.000

4.5.3 Experimental result for QUES system

The comprehensive experimental analysis done by using QUES dataset is presented in Table 4.5.3. The object-oriented metrics used in this study are DIT, MPC, RFC, LCOM, DAC, WMC, NOM, SIZE1 and SIZE2. All the validity tests described in section 4.5.1 are performed to measure the system validity.

Table 4.5.3: QUES Experimental result

Validity tests	**MLP MODEL**
R Square	0.9880
r-Correlation Coefficient	0.976
r-Square	0.9525
Mean Square Error	89.643

Root Mean Square Error	9.4680
Mean absolute Error	5.264
Min absolute Error	0
Max absolute Error	42.267
p-Value	0.000

4.6 SUMMARY

This chapter presents a brief overview of the SPSS software and its usefulness. The snapshots obtained after implementation of this study are displayed by using SPSS output viewer. The comprehensive experimental analysis, that is done by using UIMS and QUES dataset shows that the proposed model is quite valuable for real life applications. As the R Square values for both data sets are nearly 1, so this MLP model comes under a very good fit category. It signifies that the proposed model is very useful and valid. Relative performance in comparison to previous approaches is described in subsequent chapters.

Chapter 5

PRESENTATION OF RESULTS AND THEIR ANALYSIS

5.1 INTRODUCTION

This chapter outlines the presentation of results and their analysis. The goodness of fit test is carried out and the prediction results of the proposed MLP model are compared with the existing models such as Ward neural network and General Regression Neural Network (GRNN) neural network model proposed by Thwin *et al.* [97].

5.2 GOODNESS OF FIT ANALYSIS

The goodness of fit of the model is measured by using the coefficient of multiple determination (R-square), the coefficient of correlation(r), r-square, mean square error, mean absolute error, minimum absolute error, and maximum absolute error. Co-efficient correlation is represented by (r), which finds the existing correlations between the actual change and the predicted change. These statistical measures are illustrated in section 4.5.1. All these analyses are carried out to measure the model's performance.

5.3 COMPARISON OF RESULTS WITH OTHER MODELS

From the previous study, two neural network models, namely Ward neural network and General regression Neural Network (GRNN) were chosen for comparison [97].

5.3.1 Comparison of UIMS System Result

In UIMS system, r (correlation coefficient) value is 0.7798 in Ward neural network and 0.3547 in GRNN, whereas in MLP model it is 0.946. It shows that MLP model is better than Ward neural network model and GRNN neural network, which was proposed by Thwin and Quah. The significance level of a cross validation is indicated by a p value. A universally accepted p value is 0.05. In the UIMS system, mean absolute error value is 31.908 in Ward neural network and 0.300 in GRNN network whereas in MLP model it is 0.023. A commonly accepted p value is 0.05. A two-tailed probability p-value of 0.000 in all cross-validation shows a high degree of confidence for the successful validations. We conclude that the MLP model prediction is valid and useful in the population. Table 5.3.1 shows experimental validity.

Table 5.3.1: UIMS Results Comparison

Validity tests	Ward [97]	GRNN [97]	MLP MODEL
R Square	0.5444	0.3547	0.8274
r-Correlation Coefficient	0.7798	0.6984	0.946
r-Square	0.6081	0.4877	0.8949
Mean Square Error	2294.174	3249.564	532.565
Mean absolute Error	31.908	29.259	17.860
Min absolute Error	1.567	0.300	0.023
Max absolute Error	204.171	266.249	67.104
p-Value	0.000	0.000	0.000

5.3.2 Comparison of QUES system result

In QUES system, r (correlation coefficient) value is 0.747 in Ward neural network and 0.8590 in GRNN whereas in MLP model it is 0.976. It shows that MLP model is better than Ward neural network model and GRNN neural network, which was proposed by Thwin and Quah. The significance level of a cross validation is indicated by a p value. A universally accepted p value is 0.05. In QUES system, mean absolute error value is 20.782 in Ward neural network and 12.182 in GRNN network whereas in MLP model it is 42.267. A commonly accepted p value is 0.05. A two-tailed probability p-value of 0.000 in all cross-validation shows a high degree of confidence for the successful validations. We conclude that the impact of MLP model prediction is valid and useful in the population. Table 5.3.2 shows experimental validity.

Table 5.3.2: QUES Results Comparison

Validity tests	Ward [97]	GRNN [97]	MLP MODEL
R Square	0.5545	0.7220	0.9880
r-Correlation Coefficient	0.747	0.8590	0.976
r-Square	0.558	0.7379	0.9525

Mean Square Error	817.004	509.790	89.643
Mean absolute Error	20.782	12.182	5.264
Min absolute Error	0.094	0	0
Max absolute Error	114.161	109.385	42.267
p-Value	0.000	0.000	0.000

5.4 SUMMARY

Thwin *et al.* [97] found that GRNN network model can predict more accurately than Ward neural network. However, after all the experimental analysis, it is concluded that the MLP model's predictive ability is much better compared to both the models.

Chapter 6

CONCLUSION AND PROPOSAL FOR FUTURE WORK

6.1 CONCLUSION

The major purpose of this dissertation is to estimate maintainability with the help of designing metrics linked to the quality of software in object-oriented systems. This study predicted the maintenance effort using Multi-Layer Perceptron neural network model. The object-oriented metrics used in this study are valuable in estimating software maintainability. The Multi-Layer Perceptron Network model is found to be more appropriate for prediction purposes than other existing models proposed so far, as observed from the results presented above. Thus, Artificial Neural Networks can predict software in a valid and useful manner.

6.3 FUTURE WORK AND LIMITATIONS

As the size of both UIMS and QUES datasets is small, results achieved for the proposed MLP neural network model may be of limited potential. Conversely, results obtained from the current study are relatively informative and furnish the expectation that it will be appropriate for influential applications.

In the proposed work, we only considered the MLP model, whereas other neural networks, including Radial Basis Function (RBF), may also be used for the evaluation of maintainability.

There are also other emerging techniques evolving in software engineering like Neuro-Fuzzy, which can also be used for estimating these facets, as integrating the fusion approach may be beneficial. As a result, the future work in this direction needs to be carried out to further investigate a variety of emerging approaches.

In the future, we also plan to move forward with our study on other huge data sets and industrial object-oriented software systems. This research field will explore further to minimize the issues associated with software maintainability and object-oriented software systems.

REFERENCE

[1] M. R. Lyu, Handbook of software Reliability Engineering, IEEE Computer Society Press, McGraw Hill, 1996.

[2] R. E. Johnson and B. Foote Designing Reusable Classes. Journal of Object-Oriented Programming, vol. 1, no. 2, pp. 22-35, 1988.

[3] B.cox, Message/Object Programming: An Evolutionary Change in Programming Technology, IEEE Software, 1:1, pp. 50-61, Jan. 1984

[4] P. Wegner, "Dimensions of Object-Based Language Design," Tech. Report No. CS-87-14, Brown University, 1987.

[5] J.A. Mccall, P.K. Richards, and G.F. Walters, Factors in software quality, Vols I-III, Rome Air Development Centre, Italy, 1977

[6] T. DeMarco Controlling Software Projects; Management, Measurement and Estimation. Yourdan Press, New Jersey, 1982.

[7] B. Boehm, Software Engineering Economics, Englewood Cliffs, NJ: Prentice-Hall, ISBN 0-13-822122-7, 1981.

[8] E. Burt Swanson, The dimensions of maintenance. Proceedings of the 2nd international conference on Software engineering, San Francisco, 1976, pp. 492 — 497.

[9] ISO/IEC 14764:2006 Software Engineering — Software Life Cycle Processes — Maintenance.

[10] S. R. Schach, B. Jin, D.R. Wright, G.Z. Heller, J. Offutt Quality Impacts of Clandestine Common Coupling, Software Quality Journal, 11, 2003, pp. 211-218.

[11] IEEE, IEEE Standard Glossary of Software Engineering Terminology, report IEEE Std 610.121990, IEEE, 1990.

[12] ISO/IEC, Information technology - Software product quality - Part 1: Quality model, report ISO/IEC FDIS 9126-1:2001 (E), ISO, 2001.

[13] IEEE, "IEEE Standard for Software Maintenance", IEEE Std. 1219, The Institute of Electrical and Electronics Engineers, 1998.

[14] P. Goodman, "Practical Implementation of Software Metrics", McGraw Hill, London, 1993.

[15] R. S. Pressman, Software engineering: a practitioner's approach, New York: McGraw-Hill, 1982.

[16] M. H. Halstead, Elements of Software Science, ser. Operating, and Programming Systems. New York, NY: Elsevier, vol. 7, 1977.

[17] T. J. McCabe, "A complexity measure." IEEE Trans. Software Eng., vol. 2, no. 4, 1976, pp. 308–320.

[18] P. W. Oman and J. R. Hagemeister, "Construction and testing of polynomials predicting software maintainability." Journal of Systems and Software, vol. 24, no. 3, 1994, pp. 251–266.

[19] D. M. Coleman, D. Ash, B. Lowther, and P. W. Oman, "Using metrics to evaluate software system maintainability." IEEE Computer, vol. 27, no. 8, 1994, pp. 44–49.

[20] (IEEE 982.1-88) IEEE Std 982.1-1988, IEEE Standard Dictionary of Measures to Produce Reliable Software, 1988.

[21] W.S. McCulloch and W. Pitts. A Logical Calculus of the Ideas Immanent in Nervous Activity. Bulletin of Mathematical Biophysics, Vol 5, pp 115-133, 1943. Reprinted in Anderson & Rosenfeld, pp 18-28, 1988.

[22] D.O. Hebb, The organization of behaviour – a neurophysiological theory. Wiley, New York, 1949.

[23] F. Rosenblatt, The Perception: A Probabilistic Model for Information Storage and Organization in the Brain, Psychological Review, ISSN 0033-295X, Vol. 65, No. 6, pp. 386-408, 1958.

[24] B. Widrow and M.E. Hoff "Adaptive Switching Circuits," IRE WESCON Convention Record, New York, pp. 96-104, 1960.

[25] D. E. Rumelhart, and J. L. McClelland, Parallel Distributed Processing: Explorations in the Microstructure of Cognition, Volume 1 (Foundations), The MIT Press, ISBN 0-262-68053-x, Cambridge, Massachusetts, 1986.

[26] B. Cheng and D.M. Titterington Neural networks: A review from a statistical perspective. Statistical Science, vol 9, pp.2-54. 1994

[27] B. Warner and M. Misra, "Understanding neural networks as statistical tools", The American Statistician, vol 50, issue 4, pp.284-293, 1996.

[28] G.P. Zhang, E.P. Patuwo and M.Y. Hu, "Forecasting with artificial neural networks: The state of the art", International Journal of Forecasting, vol 14, pp. 35-62, 1998.

[29] F. Nielsen, Neural Networks – algorithms and applications. 2001 Available at: http://www.glyn.dk/download/Synopsis.pdf. Accessed on Dec.2011.

[30] Available: at: http://www.gc.ssr.upm.es/inves/neural/ann1/concepts/concepts.htm. Accessed on Dec.2011.

[31] Available: Wikipedia (2011), www. wikipedia.org, 2011

[32] A. Idri, T.M. Khoshgoftaar, A. Abran, Can neural networks be easily interpreted in software cost estimation?, in Proceedings of IEEE International Conference on Fuzzy Systems, pp.1162–1167. 2002.

[33] P.V.G.D Prasad Reddy, K.R. Sudha, S. P. Rama and S.N.S.V.S.C Ramesh, "Software Effort Estimation using Radial Basis and Generalized Regression NeuralNetworks", Journal Of Computing, Volume 2, Issue 5, May 2010

[34] W. Li and S. Henry, "Object-Oriented Metrics that Predict Maintainability", Journal of Systems and Software, vol 23, no.2, pp.111-122, 1993.

[35] H. D. Rombach, "A controlled experiment on the impact of software structure on maintainability", Software Engineering, IEEE Transactions on, SE-13(3):344–354, March 1987.

[36] D. R. Moreau, "A Programming Environment Evaluation Methodology for Object-Oriented Systems", Ph.D. Dissertation, University of Southwestern Louisiana, Sep. 1987.

[37] D. R Moreau and W. D. Dominick, "Object-Oriented Graphical Information Systems: Research Plan and Evaluation Metrics," Journal of Systems and Software, vol. 10, 1989, pp. 23-28.

[38] D. R. Moreau and W. D. Dominick, "A programming environment evaluation methodology for object oriented systems: part I - the methodology," Journal of Object-Oriented Programming, vol. 3, May/Jun. 1990, pp. 38-52.

[39] D. Mancl and W. Havanas, "A Study of the Impact of C++ on Software Maintenance", Conference on Software Maintenance, San Diego, CA, USA, IEEE Computer Society Press, 1990.

[40] N. Wilde and R. Huitt, "Maintenance support for object-oriented programs", Software Engineering, IEEE Transactions, vol 18, no. 12, 1992, pp. 1038 – 1044.

[41] S. R. Chidamber and C. F. Kemerer ,"Towards a Metrics Suite for Object Oriented design". Proc. Conference on Object-Oriented Programming: Systems, Languages and Applications (OOPSLA'91), Published in SIGPLAN Notices, vol 26 no. 11, 1991, pp.197-211.

[42] S. R. Chidamber and C. F. Kemerer, "A metrics suite for object oriented design." IEEE Trans. Software Eng., vol. 20, no. 6, 1994, pp. 476–493.

[43] S. R. Chidamber, D. Darcy and C. F. Kemerer, "Managerial use of Metrics for Object-Oriented Software: An Exploratory Analysis", IEEE Transactions on Software Engineering, vol.24 no.8, 1998, pp. 629-639.

[44] W. Li and S. Henry, "Object-Oriented Metrics that Predict Maintainability", Journal of Systems and Software, vol 23, no.2, 1993, pp.111-122.

[45] V. Basili, L. Briand and W. Melo, "A Validation of Object-Oriented Design Metrics as Quality Indicators", IEEE Transactions on Software Engineering, vol. 22, no.10, 1996, pp. 751-761.

[46] B. F. Abreu, and W.L. Melo "Evaluating the Impact of Object-Oriented Design on Software Quality", Third International Software Metrics Symposium, Berlin, Germany, March 1996.

[47] R. Harrison, L. Samaraweera, M. Dobie, P. Lewis, "An Evaluation of Code Metrics for Object-Oriented Programs," Information and Software Technology, vol 38, 1996, pp.443-450.

[48] Binkley and S. Schach, "Validation of the Coupling Dependency Metric as a risk Predictor", Proceedings in ICSE 98, 1998, pp. 452-455.

[49] M.H. Tang, M.H. Kao, and M.H. Chen, "An Empirical Study on Object Oriented Metrics," Proc. Sixth Int'l Software Metrics Symp., 1999, pp. 242-249.

[50] S. Muthanna, K. Kontogiannis, K. Ponnambalaml and B. Stacey, "A Maintainability Model for Industrial Software Systems Using Design Level Metrics", In Working Conference on Reverse Engineering (WCRE'00), 2000.

[51] M. Polo, M. Piattini and F. Ruiz, "Using code metrics to predict maintenance of legacy programs: a case study", 2001.

[52] M. Genero, M. Piattini, E. Manso, G. Cantone, "Building UML class diagram maintainability prediction models based on early metrics", Proceedings 5th International Workshop on Enterprise Networking and Computing in Healthcare Industry, , IEEE, 2003, pp. 263-275.

[53] R. Subramanyam and M. S. Krishnan, "Empirical Analysis of CK metricsfor Object Oriented Design Complexity: Implications of Software defects" IEEE transactions on Software Engineering, vol 29, no 4, 2003.

[54] J.H. Hayes, S.C. Patel and L. Zhao, "A Metrics-Based Software Maintenance Effort Model," Proc. 8th European Conference on Software Maintenance and Reengineering (CSMR'04), 24 – 26 Mar. 2004, IEEE Computer Society, 2004, pp. 254 – 258.

[55] M. Kiewkanya, N. Jindaswat and P. Muenchaisri, "A Methodology for constructing Maintainability Model of Object Oriented Design", Proc. 4th International Conferences on Quality Software", IEEE Computer 8-9 Sept, Society, 2004, pp.206-213.

[56] Alain, J. H. Hayes, A. Abran and R. Dumke, "Software Maintenance Maturity Model (SMmm)-The software maintenance process model", Journal of Software Maintenance and Evolution Research and Practice, vol. 17, no. 3, 2005, pp. 197-223.

[57] M. Bocco, D. Moody and M. Piattini, "Assesing the capability of internal metrics as early indicators of maintenance effort through experimentation", Journal of system maintenance and evolution, vol.17, 2005, pp. 225-246.
[58] M. Genero, E. Manso, A. Visaggio, G. Canfora and M. Piattini Building measure-based prediction models for UML class diagram maintainability, Published online: Springer Science, 21 March 2007
[59] K. M. Breesam, "Metrics for Object Oriented design focusing on class Inheritance metrics", 2nd International conference on dependability of computer system IEEE, 2007.
[60] C. Neelamegan and M. Punithavalli, "A Survey- Object Oriented quality metrics", Global journal of Computer Sc. And Technology, Vol 9, no 4, 2009.
[61] B.R. Sastry and M. V. V. Saradhi "Impact of software metrics on Object Oriented Software Development life cycle", International Journal of Engineering Science and Technology, Vol 2, no.2, 2010, pp. 67-76.
[62] Amjan Shaik, C. R. K. Reddy, A. Damodaram, "Statistical Analysis for Object Oriented Design Software Security Metrics" International Journal of Engineering Science and Technology Vol. 2, no.5, 2010, pp. 1136-1142.
[63] S.W.A. Rizvi and R.A. Khan, "Maintainability Estimation Model for Object-Oriented Software in Design Phase (MEMOOD)", Journal of Computing, Volume 2, Issue 4, April 2010.
[64] M. P. Thapaliyal and G. Verma, "Software Defects and Object Oriented Metrics–An Empirical Analysis", International Journal of Computer Applications Volume 9– No.5, November 2010, pp. 0975 – 8887.
[65] S.K. Dubey and A. Rana, "A comprehensive assessment of object oriented software system using metrics approach", International journal of computer science and engineering (IJCSE), 2010, pp. 2726-2730.
[66] C. Gautam and S.S. kang, "Comparison and Implementation of Compound MEMOOD MODEL and MEMOOD MODEL", International journal of computer science and information technologies, 2011, pp. 2394-2398.
[67] R. Malhotra and A. Jain, "Software fault prediction for object oriented System: A Literature Review", ACM SIGSOFT Software Engineering Notes, 2011.
[68] Shaik, H. Bhadriraju, K. Vikram, N. Shaik and S.V.A. Rao, "A Suggestive evaluation of system test cases in OO system trough carving and replaying differential unit test cases: A Metric Approach", International journal of Computer Science and Technology, 201, pp. 1345-1353.
[69] T.M. Khoshgaftaar, E.D. Allen, J.P. Hudepohl and S.J. Aud "Application of neural networks to software quality modeling of a very large telecommunications system," IEEE Transactions on Neural Networks, Vol. 8, No. 4, pp. 902--909, 1997.
[70] N. E. Fenton, and M. Neil, (1999), "A Critique of Software Defect Prediction Models", Bellini, I. Bruno, P. Nesi, D. Rogai, University of Florence, IEEE Trans. Softw. Engineering, vol. 25, Issue no. 5, pp. 675-689.
[71] D. Giovanni (2000), "Estimating Software Fault-Proneness for Tuning Testing Activities" Proceedings of the 22nd International Conference on Software Engineering (ICSE2000), Limerick, Ireland, Jun.2000.

[72] T. M. Khoshgoftaar, E. B. Allen, Zhiwei Xu, "Predicting testability of program modules using a neural network", In Proc. of 3rd IEEE Symposium on Application-Specific Systems and Software Engineering Technology, pp.57-62, 2000.

[73] E.L. Emam, W. Melo and C.M. Javam "The Prediction of Faulty Classes Using Object-Oriented Design Metrics", Journal of Systems and Software, Elsevier Science, pp. 63-75, 2001.

[74] F. Fioravanti and P. Nesi "A study on fault-proneness detection of object-oriented systems", Fifth European Conference on Software Maintenance and Reengineering, pp. 121 –130, 2001.

[75] M. Genero, J. Olivas, M. Piattini and F. Romero ""Using metrics to predict OO information systems maintainability", Proceedings. of the 13th International Conference Advanced Information Systems Engineering, Interlaken, Switzerland, 2001.

[76] L. Briand, W.L. Melo and J. Wust "Assessing the applicability of fault-proneness models across object-oriented software projects", IEEE Transactions on Software Engineering, vol. 28 pp. 706 –720, 2002.

[77] J. T. S. Quah, M. M. T. Thwin, "Prediction of Software Readiness Using Neural Network", In Proceedings of 1st International Conference on Information Technology & Applications, Bathurst, Australia, pp. 2312-2316, 2002.

[78] P. Yu, T. Systa and H. Muller, "Predicting fault proneness using OO metrics. An industrial case study", Proceedings. of 6th European Conference on Software Maintenance and Reengineering, pp. 99 –107, 2002.

[79] T. Menzies, K. Ammar, A. Nikora, and S. Stefano, "How Simple is Software Defect Prediction?", Journal of Empirical Software Engineering, Oct.2003.

[80] A. Mahaweerawat, "Fault-Prediction in object oriented software's using neural network techniques", Advanced Virtua and Intelligent Computing Center (AVIC), Department of Mathematics, Faculty of Science, Chulalongkorn University, Bangkok, Thailand, pp.1-8, 2004.

[81] T. Gyimothy, R. Ferenc and I. Siket, "Empirical validation of object oriented metrics on open source software for fault prediction", IEEE Trans. Software Engineering, vol. 31, Issue 10, pp.897 – 910, Oct.2005.

[82] P. Bellini, "Comparing Fault-Proneness Estimation Models", 10th IEEE International Conference on Engineering of Complex Computer Systems (ICECCS'05), pp. 205-214, 2005.

[83] L. Tian and A. Noore, "Evolutionary neural network modelling for software cumulative fault prediction", Reliability Engineering & system safety, vol. 87, pp. 45-51, 2005.

[84] M. M. T. Thwin,T. S. Quah, "Application of neural networks for software quality prediction using Object-oriented metrics", Journal of systems and software, Vol.76, No.2, pp.147-156, 2005.

[85] Y. Ma, L. Guo, "A Statistical Framework for the Prediction of Fault-Proneness", West Virginia University, Morgantown, 2006.

[86] Sandhu, Parvinder Singh, Sunil Kumar and Hardeep Singh, "Intelligence System for Software Maintenance Severity Prediction", Journal of Computer Science, Vol. 3, issue 5, pp. 281-288, 2007

[87] Q. Hu and C. Zhong, "Model of predicting software module risk based on neural network"(in Chinese), Computer Engineering and Applications, Vol.43, No.18, pp.106-110, 2007.
[88] Y. Zhao, C. Zhong, Z. Li, T. Yan, "Object-Oriented Software Fault-Proneness Prediction Using Support Vector Machine" (in Chinese), Computer Engineering & Science, Vol.30, No.11, pp.115-117, 2008.
[89] Y. Singh, A. Kaur, R. Malhotra," Predicting Testing Effort using Artificial Neural Network," Proceedings of the World Congress on Engineering and Computer Science, WCECS, San Francisco, USA, pp. 22 - 24, Oct.2008.
[90] P.S. Sandhu, U. Malhotra, and E. Ardil, "Predicting the Impact of the Defect on the Overall Environment in Function Based Systems", World Academy of Science, Engineering and Technology, 2009
[91] A. Kaur, K. Kaur and R. Malhotra, "Soft Computing Approaches for Prediction of Software Maintenance Effort," International Journal of Computer Applications, Vol. 1, no.16, 2010.
[92] R. Ratra, N.S. Randhawa, P. Kaur, G. Singh," Early Prediction of Fault Prone Modules using Clustering Based vs. Neural Network Approach in Software Systems," IJECT Vol. 2, Issue 4, Oct . –Dec. 2011.
[93] Y. Dash, S.K. Dubey and A. Rana, "Maintainability Measurement in Object Oriented Paradigm", International Journal of Advanced Research in Computer Science (IJARCS), Vol.3, no.2, pp. 207-213, April 2012.
[94] K. K. Aggarwal, Y. Singh,A. Kaur and R. Malhotra, "Application of Artificial Neural Network for Predicting Maintainability using Object-Oriented Metrics, World Academy of Science, pp. 140-144, 2006.
[95] Y. Dash and S.K. Dubey, "Quality Prediction in Object Oriented System by Using ANN: A Brief Survey", International Journal of Advanced Research in Computer Science and Software Engineering (IJARCSSE), Vol. 2, no. 2, Feb. 2012.
[96] T.M. Khoshgaftaar and R.M. Szabo, "Improving Neural Network Predictions of Software Quality Using Principal Components Analysis", IEEE, pp. 3295-3300, 1994.
[97] M. M. T. Thwin,T. S. Quah, "Application of neural networks for software maintainability prediction using Object-oriented metrics", Journal of systems and software, Vol.76, No.2, pp.147-156, 2005.
[98] K. K. Aggarwal, Y. Singh, A. Kaur and R. Malhotra, "Investigating effect of design metrics on fault proneness of object oriented systems, Journal of Object Technology,vol.6, no. 10, pp. 127-141, 2007.
[99] A. Shaik, N. Satyanarayana, M Huzaifa, N. Shaik, M. Z. Naveed, S. V. A. Rao and C. R. K. Reddy. "Invetsigate the result of object oriented design software metrics on fault proneness in object oriented systems: A case study," Journal of emerging trends in computing and emerging sciences, Vol. 2, no. 4, pp. 201-208, 2011.
[100] C. Zhong, Q. Hu, F. Yang and M. Yin. "Software Quality Prediction Method with Hybrid Applying Principal Components Analysis and Wavelet Neural Network and Genetic Algorithm,' International Journal of Digital Content Technology and its Applications, Vol. 5, no. 3, pp.225-234, 2011.
[101] F. Lanubile, A. Lonigro, G. Visaggio, "Comparing models for identifying fault-prone software components", In: Proc. of the 7th Int'l. Conf. Software Eng. and Knowledge Eng., pp. 312–319, June 1995.

[102] L.C. Briand, J. Wüst, and H. Lounis, "Replicated Case Studies for Investigating Quality Factors in Object-Oriented Designs," Empirical Software Engineering. International Journal (Toronto, Ont.), 6(1), pp.11–58. 2001.

[103] N. Katiyar and Y. H. Siddiqui, "Estimation of Software Quality using Object-Oriented Metrics through Artificial Neural Network," VSRD-TNTJ, Vol. I (3), pp.110-118, 2010.

[104] D. E. Rummelhart, G. E. Hinton, and R. J. Williams, "Learning representations by back-propagating errors," Nature, 323, pp.533–536, 1986.

[105] Available [online], http://en.wikipedia.org/wiki/SPSS.

[106] Available [online], http://www-01.ibm.com/software/analytics/spss/

[107] C.R. Kothari, Research Methodology- Methods and Techniques, New Delhi, Wiley Eastern Limited, 1985.

[108] R. Kumar, Research Methodology-A Step-by-Step Guide for Beginners, (2nd.ed.),Singapore, Pearson Education, 2005.

More Books!

yes

I want morebooks!

Buy your books fast and straightforward online - at one of world's fastest growing online book stores! Environmentally sound due to Print-on-Demand technologies.

Buy your books online at
www.morebooks.shop

Kaufen Sie Ihre Bücher schnell und unkompliziert online – auf einer der am schnellsten wachsenden Buchhandelsplattformen weltweit! Dank Print-On-Demand umwelt- und ressourcenschonend produzi ert.

Bücher schneller online kaufen
www.morebooks.shop

info@omniscriptum.com
www.omniscriptum.com

OMNIScriptum

FSC
www.fsc.org
MIX
Papier | Fördert gute Waldnutzung
FSC® C083411

Kolibri 360 GmbH
Ferdinand-Jühlke-Straße 7
99095 Erfurt, Deutschland
produktsicherheit@kolibri360.de

Druck:
CPI Druckdienstleistungen GmbH
im Auftrag der
Kolibri 360 GmbH
Ferdinand-Jühlke-Str. 7
99095 Erfurt